D1523286

SAFETY AND SURVIVAL ON THE FIREGROUND

SAFETY AND SURVIVAL ON THE FIREGROUND

Vincent Dunn

FIRE ENGINEERING BOOKS & VIDEOS

LIBRARY OF CONGRESS CATALOGING-IN-PUBLICATION DATA

Dunn, Vincent,
Safety and survival on the fireground / Vincent Dunn.
p. cm.
ISBN 0-912212-23-3
1. Fire extinction—Safety measures. I. Title.
TH9182.D87 1992
628.9'25—dc20 91-42736

Printed in the United States of America

Book design by Bernard Schleifer

9 8 7 6

To Pat

This book is dedicated to the
one hundred and twenty firefighters of America
who will die in the line of duty next year

Contents

Foreword

VINCENT DUNN'S CONSCIENCE is clear. He has gotten it off his chest. He no longer carries around with him the terrible burden of knowing all those fireground dangers and safe firefighting practices that could save a firefighter's life. Now they also appear in this book. Dunn no longer needs to worry about the possibility that a young firefighter might die or be seriously injured because he was not aware of a firefighting hazard or of a safe way of accomplishing a firefighting tactic. Every fireground danger that Vinnie has experienced, investigated, researched, and studied is written down in this book. This book is his shouted "Watch out!" to the 100 to 130 firefighters who are about to die in the line of duty next year and to the 100,000 other firefighters who will be injured next year. Every fireground hazard, danger, risk, peril, and threat imaginable is described here. And, for the first time, an attempt to categorize firefighting tactics according to their relative dangers has been set down. Of course, all of this would have fallen short of the need of the fire service if Dunn did not also include in this book the safe firefighting practices that can help all of us survive the risks of what has often been described as the most hazardous occupation in America—firefighting.

This book must be read and re-read by every firefighter who responds to fires and emergencies, every company officer who commands a fire company, and every chief officer who is responsible for the safety of firefighters on the fireground.

Preface

HOW ARE YOU TO USE this book? Do not accept all these firefighting hazards as the most dangerous for your fire departments. This book examines 15 firefighting tactics of a typical large, Northeast urban fire department. Fire departments throughout the United States will experience different dangers. Wildfires will pose a greater danger for a West Coast fire department. Cellar fires are almost non-existent in the Southeast where there are few below-grade areas built. A 20-pound propane gas cylinder fire is no "big deal" for a Southwest fire department in the oil and gas country. And a Midwest rural fire department which responds to one-story structures will not often experience the risks of operating on the floor above a fire.

So, how should a chief or company officer use this book? You should read it; then analyze your own firefighting tactics. Record your firefighting tactics according to relative risks and frequency of occurrence. Arrange them in a scale of relative risks and dangers. Create your own fire department risk management scale. Identify the firefighter hazards and fireground dangers associated with each tactic, and then develop your own firefighter safety and survival techniques.

I would like to thank my editor, Rose Jacobowitz, for her efforts on this and my other book *Collapse of Burning Buildings;* my daughter, Faith Dunn, for the illustrations; and the men and women of the New York City Fire Department who helped me write this book.

The photographers responsible for the outstanding action photos in the book are: Warren Fuchs, Bob Athanas, Alan Simmons, Bob Stella, Matty Daly, Tim Brown, Stanley Forman, Joe Pinto, Bill Landis, Harvey Eisner, Tim Klett, and William Tweedie.

VINCENT DUNN
Deputy Chief, FDNY

About the Author

Vincent Dunn is a deputy chief serving with the New York City Fire Department—a 35-year veteran who rose through the ranks of the department: seven years a firefighter, nine years a company officer, 19 years a chief.

Attending college at night with the assistance of the G.I. Bill, he received an A.A.S., B.A., and M.A. from Queens College, City University of New York.

An adjunct professor of Manhattan College, he taught fire engineering in the civil engineering department; currently an adjunct instructor with the National Fire Academy, he developed and teaches a residence course, "Command and Control of Fire Department Major Operations."

A contributing editor with *Fire Engineering* magazine and *Firehouse* magazine, he has authored many articles on firefighting safety and survival.

He is the author of the book and video series, "Collapse of Burning Buildings."

SAFETY AND SURVIVAL ON THE FIREGROUND

1

Death and Injury Statistics

EACH YEAR more than 100 firefighters die in the line of duty, and another 100,000 are injured. This fact translates into one firefighter death every three days, and 8,000 firefighter injuries every month.

If a firefighter dies while combating a blaze, that death will be caused by the extreme physical and emotional stress of firefighting while pulling a hose line or raising a ladder, an apparatus accident while responding to or returning from alarms, falling off a smoky roof at night, being trapped by smoke and flame, or being buried alive under tons of bricks and timbers during a structural collapse.

Causes of Firefighter Deaths

The NFPA *Annual Death Statistics 1991* lists these leading causes of firefighter deaths (see Figure 1.1):

1. Stress
2. Responding, returning to alarms
3. Falls, falling objects, in contact with
4. Products of combustion
5. Collapse

Causes of Firefighter Injuries

If a firefighter is not killed but is instead seriously injured, the injury will be due to a laceration or deep puncture wound which cuts an artery, tendon, or nerve; a torn back or ankle requiring immobilization and a brace, unconsciousness from inhaling smoke or toxic fumes, or suffering a third-degree burn, or a blinding eye injury.

Fire Service Deaths 1977 to 1990

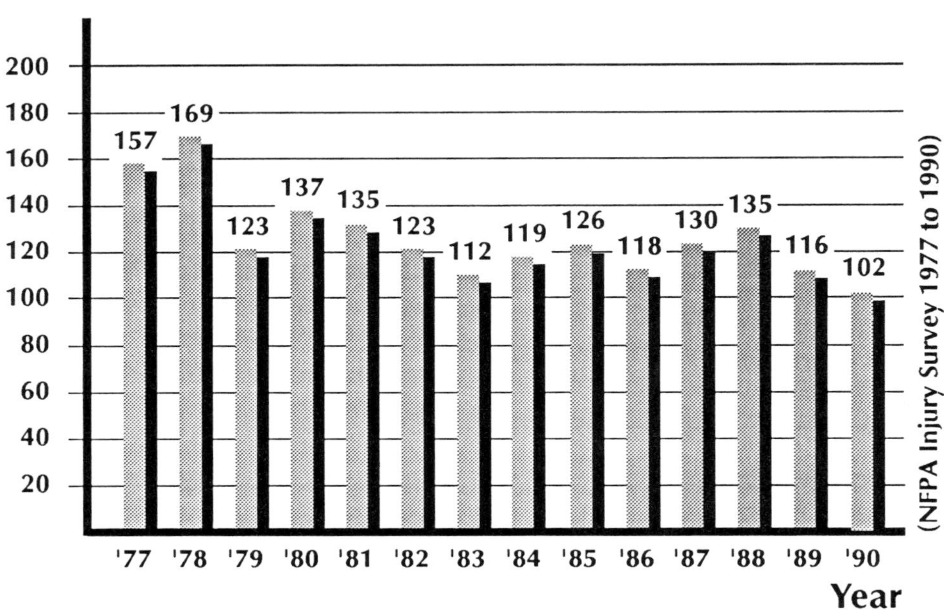

Figure 1.1. *Each year between 100 and 130 firefighters die in the line of duty.*

The NFPA *Annual Injury Statistics 1991* lists these leading causes of firefighter injures (see Figure 1.2):

1. Wounds, cuts, bruises
2. Strains and sprains
3. Smoke or gas inhalation
4. Burns
5. Eye injuries

Firefighting is a high-risk, dangerous occupation. The major risks of firefighting are created by the flames, heat, smoke, and toxic gases produced by fire. Unlike any other dangerous occupation, a firefighter must work in an extremely dangerous environment, constantly threatened by flame and smoke as he performs his tasks. In the mining industry, for example, before a worker enters a mine shaft some basic minimum safety measures are required: air is pumped into the underground tunnels, shoring and structural supports are placed to reinforce the underground excavations, and electric lighting is provided

Nature of Fireground Injury

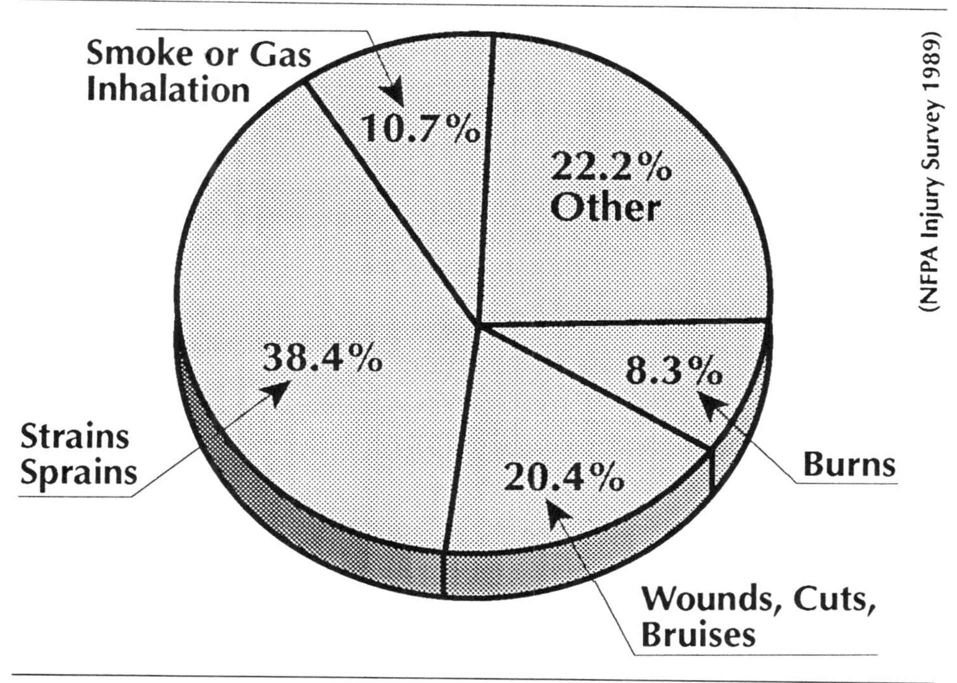

Figure 1.2. *Wounds, cuts, and bruises constitute the second leading cause of firefighter injury.*

Figure 1.3. *Protection of life is the first priority of firefighting; this priority includes the lives of firefighters.*

in the dark mine shafts and tunnels. This preparation is not the case when a firefighter goes to work. When a firefighter enters a burning building, he crawls into a flaming, smoke-filled room and enters an unpredictable, deadly work environment. There is only heat and darkness in the smoke-filled rooms; there is no breathable air in the superheated atmosphere; and the building may explode or collapse at any moment. There is no shoring or bracing in place to support the heavy plaster ceiling or the roof beams.

For the past 200 years, the dangers that firefighters were exposed to have been considered part of the job. Risks of death and injury were what they were paid for; protection of property was considered more important than the firefighters. Since the Second World War, this attitude has changed (Figure 1.3). The risks of firefighting have been reconsidered, as have the economic costs of firefighter death and injury. The profession of safety engineering was created and has grown; fire departments have faced litigation for unsafe work practices; the creation and regulations of the Occupational Safety and Health Administration have mandated change. The moral and legal priorities of firefighting strategy and tactics have been defined. Protection of life is the highest priority of firefighting, and this includes the lives of the firefighters as well as those of the people we serve. Fire containment is the second priority of firefighting. Property protection is now the last priority of firefighting strategy and tactics.

In the past decade, risk managers and safety engineers have turned the spotlight on to the workplace and the workers of America. As a result, firefighting practices and procedures have been analyzed, and fireground dangers have been identified and classified according to risk. Out of these studies have emerged new insight and understanding of firefighters' safety and survival (Figure 1.4).

There are degrees of danger in firefighting; all phases of firefighting do not present equal risks. Some firefighting tactics are much more dangerous than others. A firefighter must know the various degrees of danger he will be exposed to when he is given an assignment to carry out. He must know, for example, that raising a ladder is not as dangerous as advancing an attack hose line, that outside venting is not as dangerous as searching for the location of a fire before the hose line is in place. In order for a firefighter to survive the dangers of firefighting, he must know how other firefighters have died or been seriously injured. The list that follows gives 15 dangerous fireground tactics; firefighters have been disabled or killed performing these operations at fires. These high-risk firefighting tasks have been identified as

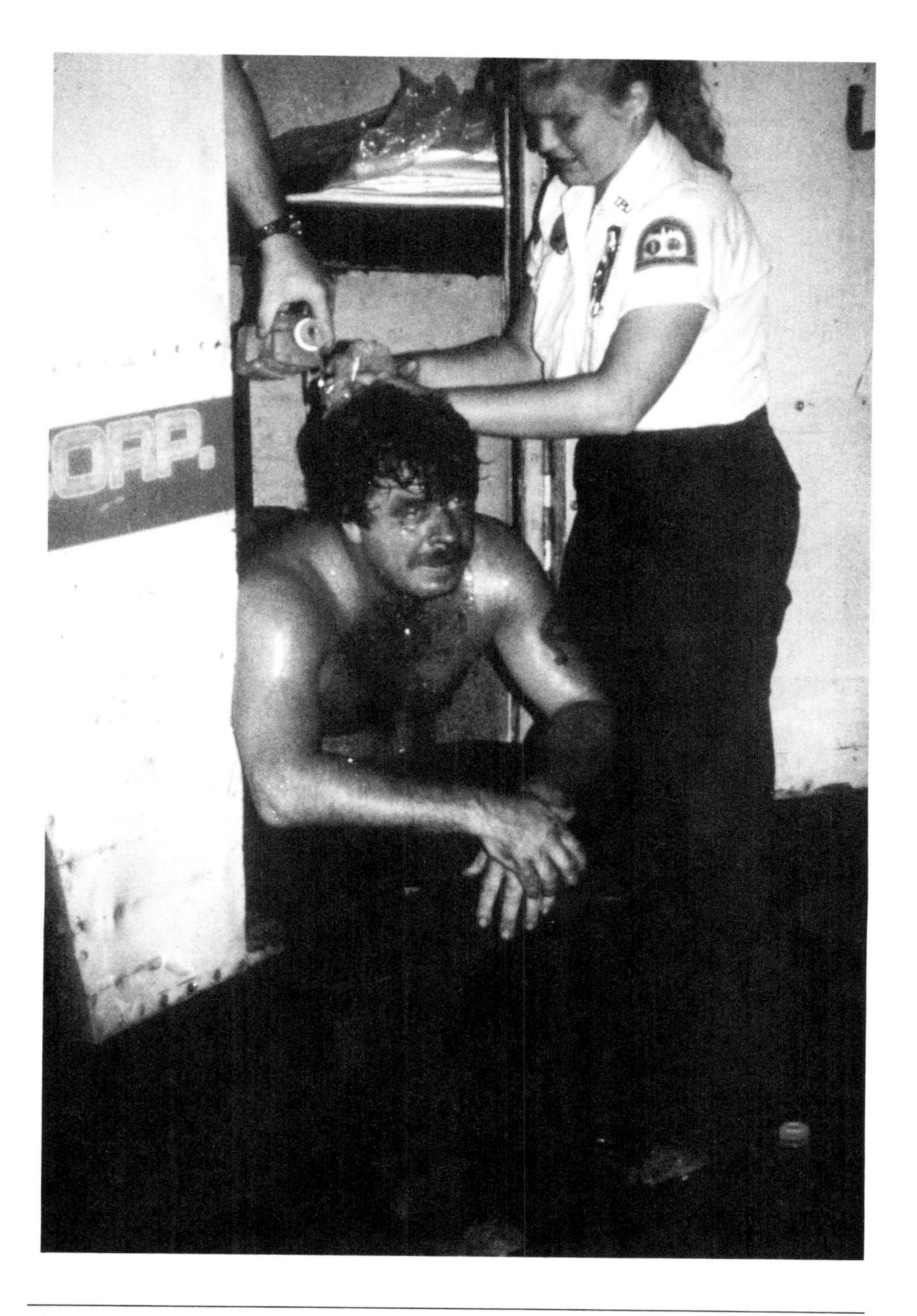

Figure 1.4. *In order for a firefighter to survive the dangers of firefighting, he must know how other firefighters have died.*

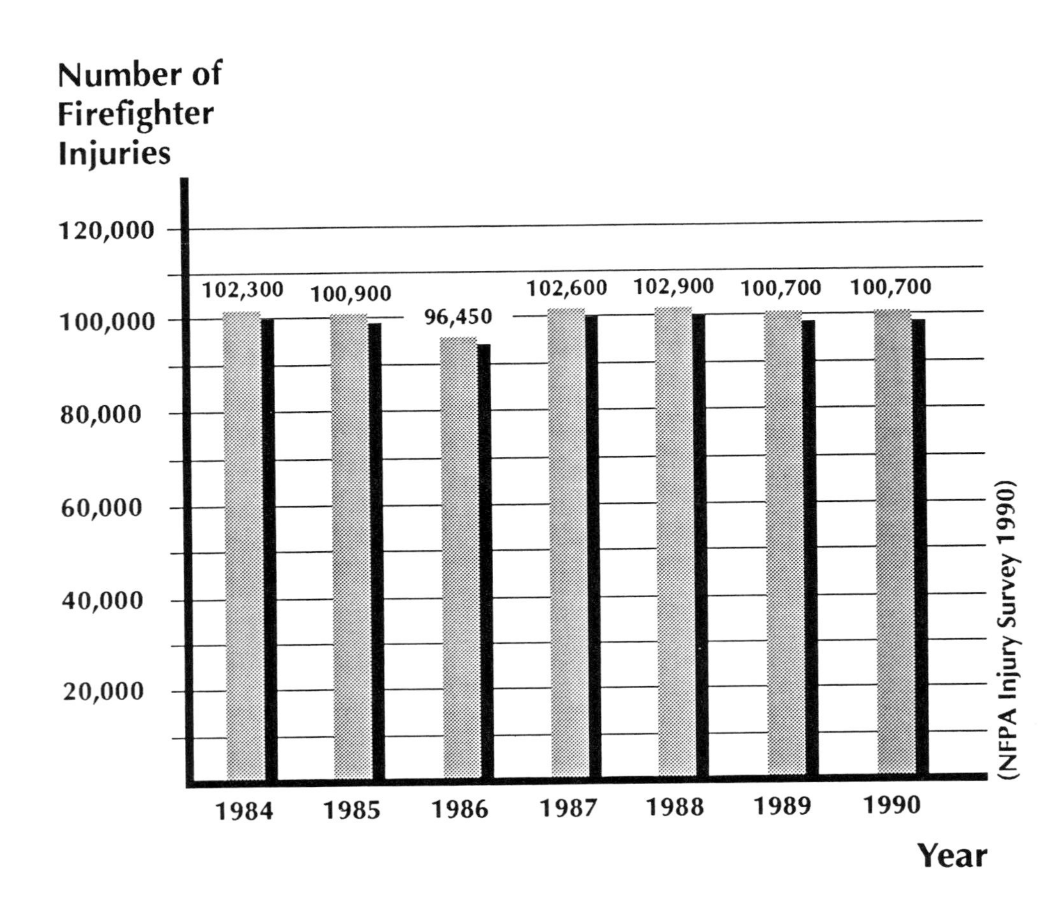

Figure 1.5. *Each year, approximately 100,000 firefighters are injured while fighting fires.*

among the most dangerous duties a firefighter can perform at a fire. The tactics were identified by studying and analyzing the United States Fire Academy records of firefighter fatalities in a 1985 pamphlet; the International Association of Firefighters, Death and Injury Annual Report in *International Firefighter* of June–July 1990; and the National Fire Protection Association, United States Firefighter Annual Death Report and Injury Report in *Fire Command*, June and November 1991 (Figure 1.5).

Dangerous Firefighting Tactics

The list of these tactics follows:

1. Collapse rescue operations
2. Responding and returning to alarms
3. Searching for a fire
4. Advancing an attack hose line
5. Operating on a peaked roof
6. Operating above a fire
7. Cellar fires
8. Propane gas fires
9. Wildfires
10. Aerial ladder operations
11. Forcible entry
12. Master stream operations
13. Outside venting
14. Fire escape operations
15. Overhauling

The 15 firefighting tactics are rated as to degree of risk on the basis of the following criteria:

1. A firefighting tactic performed before the initial attack hose line is in operation has a higher degree of risk.
2. A firefighter tactic carried out in close proximity to the flames has a higher degree of risk.
3. A firefighting tactic carried out above the fire has a higher degree of risk.
4. A firefighting tactic performed where there is a danger of explosion or collapse presents a higher degree of risk.
5. A firefighting tactic performed inside a burning building is assigned a higher degree of risk than a tactic performed outside a burning building.
6. A firefighting task performed during the early stage of a fire—the growth stage—is given a higher degree of risk than a task performed during the later stages of a fire—the fully developed or the decay stage.
7. A firefighting tactic rarely performed by firefighters is given a higher degree of risk than a tactic regularly performed by firefighters.

8. A firefighting tactic about which there is little published training information is given a higher degree of risk than a tactic for which there is an abundance of written training material.
9. A firefighting tactic which kills and injures large numbers of firefighters each year is given a higher degree of risk than a task that does not kill or injure many firefighters.
10. Firefighting tactics performed by an individual firefighter are given a higher degree of risk than a tactic performed with a team of firefighters.

How to Use the Scale of High-Risk Firefighting Tactics

There are many different ways to present information about firefighting risks, and no single graph or chart can describe a firefighting risk with absolute certainty. This scale (Figure 1.6), in spite of possible

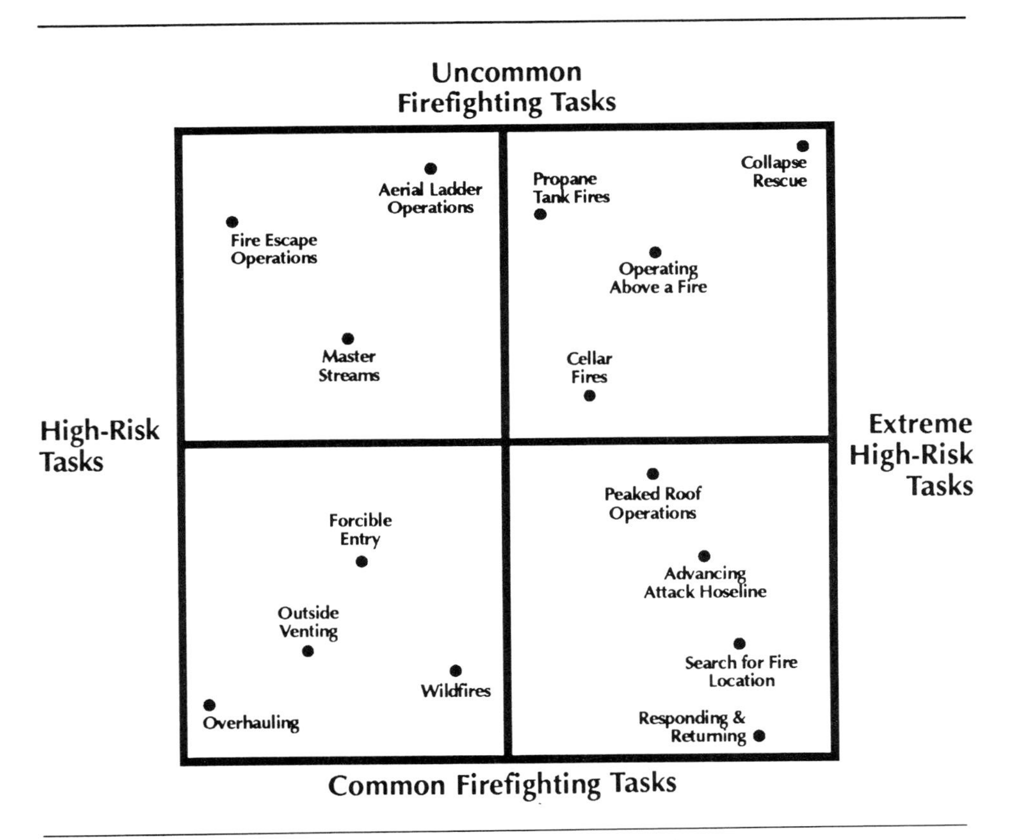

Figure 1.6. *This scale shows 15 dangerous firefighting tactics.*

overgeneralization, is intended to help us all assess more accurately the dangers of firefighting. Other firefighting tactics should be added to the scale in order to evaluate their degree of risk as compared to other firefighting tasks. The degree of risk assigned to one of the 15 firefighting tasks should be changed—moved to a higher or lower position, if necessary; new fire protection technology, equipment, or firefighting safety procedures may reduce the risks of one of the firefighting tactics, or new environmental factors, structural factors, or fuels may increase the risks of a firefighting task. It is best to remember that firefighting hazards and risks change constantly, but there is one statistic that does not: each year 100 to 130 firefighters die in the line of duty, and 100,000 others are injured.

1. *Collapse rescue operations* are considered a dangerous firefighting tactic because most collapses are accompanied by fire. Rescue operations after a building has collapsed are dangerous because there is danger of a second collapse burying rescue firefighters. Furthermore, after a collapse there is great potential for a delayed gas explosion and fire, trapping firefighters who are on top of the pile of rubble, searching for victims.
2. *Responding and returning to alarms* is not actually a firefighting tactic; however, because almost 25 percent of all firefighters killed in the line of duty each year are responding and returning to alarms, it must be considered one of the most dangerous tasks a firefighter can perform. The number of firefighters' deaths and injuries in a fire department are directly related to the number of times the department responds to alarms.
3. *Searching for the location of a fire* upon arrival at a scene is a dangerous firefighting tactic because the firefighter is operating without the protection of a hose line. The first firefighter to enter a burning building to size up a fire enters an extremely dangerous environment. Firefighters searching for the location of a fire inside a burning building have become lost in smoke, burned to death by flashover, and blown out of the building by gas explosions, boiling-liquid expanding-vapor explosions, and backdrafts.
4. *Advancing an attack hose line* is a dangerous firefighting tactic because firefighters pushing an attack hose line into a burning room must confront the fire head on. Firefighters advancing the attack hose line come in close proximity to the

raging fire and are often injured by flame, steam, burning embers, heated plaster ceiling collapse, and smoke.

5. *Operating on top of the peaked roof* of a burning house is a dangerous firefighting tactic, which has killed and injured many firefighters. Firefighters may fall while climbing a ladder up to such a roof. A firefighter might plunge through a fire-weakened roof deck into a burning attic; firefighters often lose their balance and fall from the peaked roof to the ground below.
6. *Operating on the floor directly above a fire* is a dangerous firefighting task because the fire may spread up the stairs and trap firefighters, blocking their escape downward and out of the burning building. The scientific laws of heat transfer tell us that heated air expands, becomes lighter than air, and rises. The heated convection currents of smoke, heat, and flame rise upward. Thus, the area directly above a fire is the most dangerous area of a fire building.
7. *Fighting fire in a below-grade area* is a dangerous firefighting tactic due to the inaccessibility of the fire. Access to a burning cellar is usually limited to one or two stairways. Advancing an attack hose line into a windowless cellar, down a steep stairway, is a punishing task. Heat, smoke, and flame rise up from the below-grade fire into the face masks of the firefighters who are trying to descend the steps into the cellar with the attack hose line.
8. *Propane gas fires* kill firefighters by means of exploding fireballs, flying missiles in the form of jagged pieces of metal cylinder, and shock waves produced by the boiling-liquid expanding-vapor explosion (BLEVE). Fighting fires in burning liquefied petroleum gas is a dangerous task because the flame cannot be extinguished, there is danger of an explosion, and firefighters must approach the burning cylinder closely to shut off the flow of gas.
9. *Fighting wildfires* is the second greatest killer of firefighters. After residence fires, more firefighters are killed each year battling fires in forests, brush, and grass than other types of fire. Fighting a brush fire is a dangerous task because of the explosive flammability of the vegetation serving as fuel, the unpredictable, erratic spread of fire caused by wind changes, and the effects of hills and mountains on the convection currents of large-scale wildfires.

10. *Climbing or operating on an aerial ladder* is a dangerous firefighting task. There is the constant risk of falling from a raised aerial ladder or platform. Electrocution, improper positioning, locking, or sudden movement of a raised aerial ladder can lead to a firefighter losing his grip and his balance. Safe climbing techniques and victim removal procedures are necessary skills for firefighters who operate on aerial ladders.
11. *Forcible-entry operations* are dangerous because of the firefighter's exposure to a potential backdraft explosion. When the initial entry is made into a superheated, long-smoldering, oxygen-deficient fire area, there is danger of an explosion. The closer that forcible entry is performed to the fire, the greater the risks become.
12. *A master stream* is a tremendously destructive force which delivers three or four tons of water, speeding through a nozzle at 100 feet per second. When improperly directed, a master stream can collapse a chimney, knock down a parapet wall, explode slate shingle into the air, and cave in a ceiling with tons of water. Master stream operations are a dangerous firefighting tactic because of the above reasons and also when outside streams are put into operation, there is usually a danger of burning building collapse.
13. *Outside venting* is a hazardous firefighting tactic because a firefighter must work around the perimeter of a burning building, one of the most dangerous areas of the fireground. Broken glass, building fragments, shingles, smoldering furniture, dropped tools, and people jumping from flaming windows have injured firefighters venting the outside of a burning building. Firefighters perform outside venting work alone and, in some instances, search a burning building directly in the path of an advancing attack hose line (Figure 1.7).
14. *Operating from a fire escape* is considered one of the dangerous firefighting tactics because of the possible collapse, corrosion, and deterioration of these metal and wood fire exits. Because building owners fail to repair and maintain fire escapes, it has become increasingly dangerous to climb or operate hose lines from them during fires (Figure 1.8).
15. *Overhauling* is a dangerous firefighting task. Toxic smoke and dusts fill the air of a burned-out structure after a fire. Darkness, exhaustion, uncoordinated actions, and a fire-weakened structure are major causes of firefighter injury and

Figure 1.7. *The physical and psychological stresses of firefighting are the number one killers on the fireground.*

Figure 1.8. *There are no new causes of firefighter death and injury; the only new factors are the firefighters.*

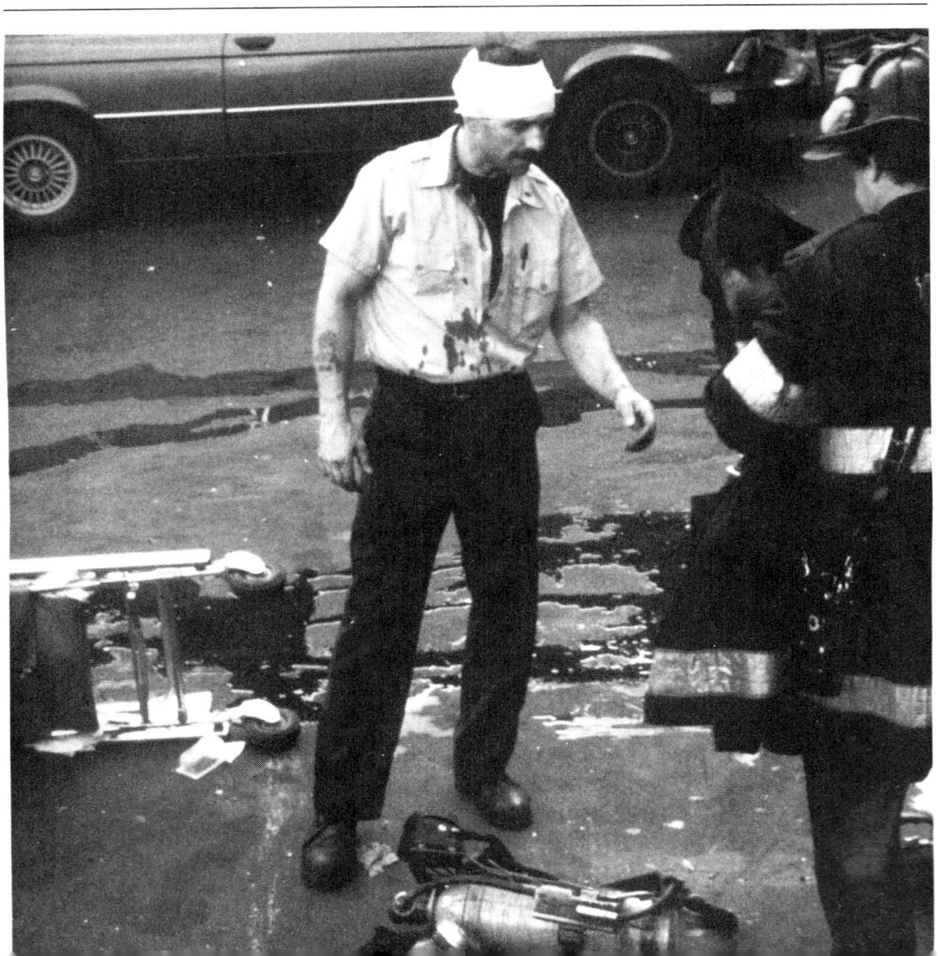

death during overhauling operations. Overhauling is a hazardous stage of a fire. Long-term chronic illness for firefighters can result from overhauling operations.

There is no firefighting tactic that does not present the danger of death or injury to a firefighter; however, some firefighting tactics do present greater risk than others. The tactics listed above are in descending order, from extremely dangerous down to dangerous. Do not be deceived into thinking that the fireground tactics at the lower end of the scale are not deadly. You can be killed or seriously injured performing any one of these firefighting operations.

There are no new causes of firefighters' death and injury; firefighters are killed and injured performing the same tasks each year. The only new factors are the firefighters. Young firefighters join the fire company, and old firefighters leave the fire company, but the types of fireground death and injury remain the same. Rookie firefighters die and become disabled in the same ways that firefighters died and were disabled 30 years ago. It is tragic to see the same fireground dangers killing and maiming firefighters year after year, in fire departments throughout the country.

Figure 1.9. *There is no fireground tactic that does not present the danger of death and injury; however, some firefighting tactics do present greater risks to firefighters than do others.*

There is no fireground tactic that does not present the danger of death or injury to a firefighter; however, some firefighting tactics do present greater risks than others. The 15 dangerous firefighting tactics are shown in a group to illustrate more clearly their varying degrees of risk and danger to firefighters (Figure 1.9).

How to Use This Book

Firefighters should not have to learn about firefighting dangers the hard way, as we did, by trial and error. Firefighters today should not have to repeat the same fireground mistakes we made. This book has been written in an attempt to reduce firefighters' death and injury, year after year, in fire department after fire department. When a rookie firefighter reports for duty at a fire station, he should not be assigned a dangerous firefighting task immediately. The new firefighter should gradually be given the more dangerous tasks only after he has shown the ability and the desire to assume more responsibility. When a firefighter is given a new assignment to perform at a fire, one that he has not carried out before, he should first examine the graph of dangerous firefighting tactics in this chapter (Figure 1.6) to determine the relative danger of the assignment and compare its danger to that of other firefighting tasks he has performed at fires. Next, the firefighter should read the chapter in this book which describes in detail the hazards and risks inherent in that particular firefighting tactic. He should also study and use the recommended safety techniques, tips, and precautions specified in the chapter to avoid suffering one of the common recurring causes of death and injury associated with that firefighting tactic. There is an old saying in the fire service: to be forewarned is to be forearmed. That forewarning is the objective of this book.

2

Collapse Rescue Dangers

"Wow! Captain, look at that building—the whole side of it collapsed!" the driver of the rescue company exclaims as the heavy-rescue truck nears the collapse scene.

The officer speaks into the apparatus radio handset. "Rescue to Communications Center, we have a major collapse of a six-story building, 25 by 85 feet, brick-and-joist construction. The exposure number four wall, floors, and roof have collapsed on top of cars in an adjacent parking lot."

Outside the apparatus, the officer of the rescue company sees the first-arriving engine company stretching a hose line to the collapse area; the ladder company is taking two people from a front window ledge—the stairway has collapsed with the floors.

Sizing up the collapse, the rescue officer looks up at the gaping hole that exposes most of the interior of the office building. The roof and the top three floors have pancaked down on the parked cars as a result of a bearing wall failure. Near the front of the collapsed building, sections of the roof and floors are hanging down at a dangerous 60-degree angle, an unsupported lean-to collapse. At the rear of the collapse, large metal shelves loaded with boxes, and loose floor timbers threaten to tumble down on the pile of rubble. Everything appears about to collapse downward on top of the area where the firefighters must search for victims (see Figure 2.1).

There is no sign of fire as yet. If, however, leaking gas from a broken gas pipe is ignited by a sparking electric wire, there will be an explosion. The firefighters must work fast.

Geared up in the street, the captain of the rescue company gives

Figure 2.1. LEFT: *A bearing wall failure caused a collapse of the six floors of this building.*

Figure 2.2. RIGHT: *A secondary collapse during a rescue operation is one of the greatest dangers faced by rescuers.*

assignments to his firefighters. "John and I are going to search the collapse pile on the exposure four side." Pointing to the remaining three firefighters in turn, he says, "You take the front of the building; you take the exposure 2 side; you take the exposure 3 side. Look for any surface victims who might be in shock or injured, and search all the voids you can find. Report back to me when you complete the search. Let's go."

The firefighters split up. The captain and the accompanying firefighter climb to the top of the pile of collapse rubble, looking for persons lying in the debris. The two men step over shelving, boxes, broken timbers, and bricks. Late afternoon sunlight casts shadows beneath sections of collapsed floors and roof. The two bend over and shine their lights into the dark spaces where bodies could be wedged. They call out into the smaller voids, "Anyone in there?"; they crawl in and out of the larger spaces. The officer looks up at the 60-foot-high, free-standing wall, at the remaining floors and shelving that could collapse at any moment (Figure 2.2).

The firefighter shines his light into a large void created by a lean-to of the third and fourth floors, and hears something. "Help," a weak voice calls out. "I'm trapped." The firefighter turns to his officer and shouts, "Say, Captain, I found someone! I'm going in to get him!"

The officer scrambles over the rubble to the place where the firefighter is tying a search rope to a timber and is about to play it out before crawling into the dark space between the two partially collapsed floors. Just then, a section of metal shelving and several cardboard boxes tumble down from the upper floors into the parking lot. The debris misses them.

Inside the void, the firefighter gropes his way toward the trapped man's voice. He lies on his back, moving sideways with a sliding motion, across desks and boxes caught between the two collapsed floors. He is at a 45-degree angle from the horizontal. The space becomes narrower. His flashlight suddenly goes out. He taps it with the hand holding the search line, and it blinks back on.

With his next movement, something suddenly gives way, and the firefighter slides down between the two floors. His feet are stuck; he is wedged between the lean-to floors at the point where they meet. He drops the rope and, grasping the floor above him, searches for a handhold to pull himself up. Finally he finds one and pulls his body up, freeing his legs.

He remembers the house of horrors at the amusement park. There, too, the walls and floors project at crazy angles—he had to brace his arms and legs against a wall just to move forward.

"Over here!" The man's voice is below him.

Scanning the area in front of him with his flashlight, the firefighter can barely make out the form of the man. The victim is lying at a 45-degree angle from the horizontal, upside down, covered with plaster dust. His legs are pinned above his head. Several crushed cardboard boxes cover his legs and feet.

"Please, get me out of here before both of us are buried!"

Quickly, the firefighter braces one leg against the floor beam above and, with both hands, starts pulling the boxes away from the man's feet. Suddenly, there is a loud crack—a section of the floor above them slides downward several inches. The firefighter freezes. He hears his own heart pounding and feels the cold sweat on his forehead.

"Hurry!" the trapped man cries. "Don't stop!" the firefighter uses a pocket knife to cut the last crushed box holding the man's leg, pulls out the stuffing from within the box, then pulls the man's leg free. The man tumbles down and then rights himself inside the small space. The firefighter drops his flashlight, which rolls out of sight. He says, "Let's go." The two men start to crawl out of the dark space.

The firefighter leads the way, hunting for the light of the void opening, but there is only darkness. The firefighter panics. "Hey, Captain, where are you?" he shouts. There is no response.

A sudden, loud crunching sound fills the dark void. The firefighter and the victim, lying side by side on their backs at a 45-degree angle from the horizontal, hands and feet wedged against the collapsed floor section above them, feel that floor section sink down several inches, reducing the space in which they are lying. "Oh, God!" the firefighter thinks to himself. "We are trapped!"

"John, John!" The voice of the captain comes from somewhere above them. "The opening is over here! See my light!" Looking up, they can see beams of light poking through the cracks in the mountain of rubble. They must somehow have slid below the opening.

The firefighter and the rescued man start to climb upward. Cardboard boxes and dust begin to fall on top of them. The firefighter climbs atop an overturned metal desk, sees the void opening, and helps the victim up on to the desk and out of the rubble (Figure 2.3).

Crawling out into the fresh air, covered with dust, the officer and the firefighter help the man off the collapse pile. "Great job," the captain says to the firefighter as they reach the sidewalk.

At 1503 hours on October 24, 1988, a six-story building collapsed. The structure, at 24 West 31st Street in New York City, was under renovation. The bearing wall on the exposure 4 side caved in, as

Figure 2.3. *Void search, along with surface search and site survey, are the most productive life-saving phases of a collapse rescue operation.*

a curtain-fall collapse, triggering the pancake collapse of the three top floors down on to cars in an adjacent parking lot. Remaining sections of the collapsed floors and roof were hanging in an unsupported lean-to configuration and could have collapsed on top of rescue workers at any moment during the entire rescue operation.

Firefighters removed two people, who were about to jump from window ledges at the front of the building, by means of an aerial platform. Police and firefighters there extricated two victims trapped inside voids. Nine hours later, firefighters dug out a woman—alive—from beneath tons of collapse rubble.

According to the American Society of Civil Engineering, there are 500 major building collapses in the United States each year. The fire service is called to most of these structural failures in order to rescue people trapped in the collapse rubble.

When a fire department responds to a collapse, the chief must have a collapse rescue plan ready to put into operation, and he must establish an incident command system to coordinate actions effectively with other agencies. Nonetheless, the individual firefighter who is ordered to climb up on the collapse rubble pile to search for and rescue trapped victims must understand the dangers that threaten him during the collapse rescue operation.

This chapter examines the dangers of a collapse rescue operation, one of the most deadly tasks a firefighter will ever be called upon to perform. All of the dangers we are about to examine were present at the collapse rescue operation which occurred in New York City on October 24, 1988.

Secondary Collapse

A secondary collapse of a partially destroyed floor or wall during a rescue operation can kill more firefighters than the victims already buried in the collapse rubble. It is the greatest danger at the collapse of a multi-story building.

Upon arrival at a building collapse, the rescue plan is: 1. Conduct a survey of all sides of the collapse for injured victims. 2. Search the rubble pile for surface victims. 3. Search all voids and spaces in the collapse rubble. 4. Remove selected debris; tunnel or trench to specific locations for buried, trapped victims. 5. Remove general debris. Steps one, two, and three of the collapse rescue plan—site survey, surface search, and void search—are the most productive. Most victims are rescued during these stages, but they are the most dangerous to the rescue workers. A second collapse of an unstable wall or floor of a multi-story building could bury firefighters beneath tons of brick and timber, or trap them inside a void they are searching.

During the initial rescue operation, the collapse structure continues to shift and settle. At collapse rescue operations thus far, the fire service has been fortunate. No firefighters have been killed during site survey, surface search, or void search. The risk/benefit (danger to firefighters/number of victims discovered) relationship of this operation is still on the side of the fire service. Although it is an extremely dangerous task for firefighters to climb up on a collapse rubble pile and search for unconscious and trapped victims, it has always proven to be the most productive life-saving action of the entire collapse rescue plan.

If, after completion of surface and void search, there is danger of a secondary collapse, all firefighters engaged in rescue work should be withdrawn from the collapse area (though fire-extinguishing efforts may have to continue if there is a spreading fire) and a conference held. A safety analysis and re-evaluation of the collapse site must be conducted by the incident commander (Figure 2.4). In the majority of cases, most of the victims will have been rescued by this time. This

Collapse Rescue Plan

1 Site Survey: Size up all 6 sides of collapsed building for secondary collapse dangers, fire or explosion dangers.

Shut Off Utilities: Gas, Electricity, Water. (Note: Most collapse plans include this action in site survey. This action is so important to collapse rescue safety that it requires equal importance and mention in the collapse rescue plan.)

2 Search the Collapse Rubble for Surface Victims.

3 Void Search: Examine all accessible spaces.

4 Selected Debris Removal: Tunnel or trench, or remove debris to gain access to specific locations for access to buried victims.

5 General Debris Removal.

Figure 2.4. *When a fire department responds to a collapse, the chief must have a collapse rescue plan.*

conference should address rescue worker safety and the precautions that will be followed to safeguard them, the names and probable locations of persons still reported missing, and the procedure for selected debris removal—tunneling or trenching.

Because the chance of survival for victims found under the rubble during the next (latter) stages of the collapse rescue plan is less, greater safety precautions should be taken to protect the firefighters and rescue workers. Unstable walls may have to be removed. Shoring of partially collapsed structures may be in order. Lights will have to be set up during nighttime rescue work. Confirmation of gas, electric, and water shut-off must be obtained.

Rescue work must not stop as long as any person is still reported missing. People have survived, trapped inside voids beneath tons of collapse rubble, for several days. (After the recent earthquake in Mexico City, an adult survived nine days and an infant eleven days.) A buried victim's chances for survival are highest if extricated from a collapse building within the first twenty-four hours. Greater safety precautions must be taken, however, to protect rescue workers during the selected debris removal operation.

Explosion

The second greatest danger to a firefighter performing collapse rescue work is explosion. When a building collapses, a gas main can rupture, gas pipes may break, and gas cylinders inside the structure are often crushed and broken. There will be leaking gas. As the gas—natural or liquified petroleum—seeps throughout the collapse rubble, it will inevitably meet a snapped or broken sparking electric wire, or a flame from some heating or cooking appliance, and explode violently. The shock waves created by this explosion can cause a secondary collapse, killing many rescue workers who are doggedly toiling on and within the collapse rubble pile to remove victims. Such an explosion will also collapse any voids where people might be trapped, who might have survived had it not been for the rapidly spreading, gas-fed fire.

Leaking gas, even if it becomes too rich a mixture to ignite with air, can displace the oxygen inside voids or in below-grade areas, killing victims and firefighters inside these confined spaces.

One of the most important assignments to be given by the officer in command of a collapse is to shut off the building services, such as gas, electricity, and water. The utility companies must be called to the scene; however, firefighters should be assigned to this task before the utility company representatives arrive. Utility companies are often delayed when responding at night or on weekends.

The gas should be the first utility shut off because it is the most dangerous during a collapse rescue operation. The first firefighters on the scene who are ordered to conduct the initial site survey, surface search, and void search must be on guard to detect the odor of leaking gas. The officer in command of the operation must be notified immediately of the presence of gas. If there is a large-diameter broken gas main and gas permeates the entire collapse site, firefighters must be withdrawn after the surface search has been completed. The void search must be delayed until the main can be shut off. Ideally, utility shut-off should be the second priority stage of the rescue operation, after site survey.

Utility shut-off takes some time to accomplish, so it should be started as soon as possible. Shut-off is usually completed during the void search. This vital procedure must be completed before tunneling or trenching operations begin.

Fire

The third greatest danger to rescue workers at a collapse site is fire. If fire occurs during a collapse rescue operation involving a brick-and-joist or wood-frame building, it will spread rapidly. Firefighters searching for victims on top of the collapse rubble pile or inside voids may be trapped by the quickly spreading flames. The fire could be fueled by flammable liquids leaking from crushed containers or machinery.

Further increasing the rate of flame spread on the rubble pile will be broken pieces of wood from the collapsed structure. After a collapse, large timbers are broken into smaller wood sections. Air spaces and voids will honeycomb the collapse rubble pile around this splintered wood. Plaster and fire-retarding surface coverings will have been broken away from wood lath sections of walls and ceilings. As the fire protection engineers say, the ratio of exposed surface area to the total volume of wood is increased manyfold by the collapse. The two factors which increase the rate of burning—air and exposed wood surface—will be present at a collapse site.

The first engine company to arrive at a collapse must stretch a hose line and protect against a fire and rapid flame spread. An aerial ladder pipe or aerial platform master stream should also be positioned near the collapse site. This aerial device should be supplied with a hose line and be ready to extinguish a fire in the collapse rubble, should that occur.

Firefighters must wear full protective clothing when conducting a surface search or void search. The chances of the collapse pile turning into a giant pyre are high. Rescue workers should be equipped with clothing to protect them from flames and heat before they participate in the collapse rescue operation. An explosion or flash fire could trap firefighters on top of the pile of broken wood (Figure 2.5). Firefighters will not be able to climb quickly down from a collapse rubble pile; the uneven surface of jagged metal, broken timbers, bricks, and holes on top of a collapse rubble pile will cause them to trip and fall.

Electricity

A building contains hundreds of feet of electric wire snaking through its walls, floors, and attic space. When a building collapses, this electric wire is threaded throughout the collapse rubble, just waiting to entangle an unsuspecting firefighter or his metal tools. Sparks from the wires can ignite small fires in nearby combustible debris.

***Figure 2.5.** Firefighters must wear full protective clothing when conducting a surface search of collapse rubble. The chances of an explosion or fire are great.*

Water pipes are also broken during a collapse, and, if there is a fire after the structural failure, fire department hose streams will be used. Water-soaked collapse rubble and exposed, broken, live electric wires, together with firefighters using metal tools to cut, pull apart debris, and dig out buried victims, create the possibility of serious injury or death by electrocution.

If the collapse structure is of lightweight steel construction, the entire structure could be electrified, preventing the most productive life-saving stages of the collapse rescue from being carried out.

Again, as with gas, the utility companies should be called to the scene to shut off the electricity; however, firefighters should be ordered by the officer in command to attempt electric supply shut-off prior to the arrival of the utility company.

If a firefighter's metal took becomes entangled in a live electric wire, the continuous electric current shooting through the metal tool and through his body may keep his hands gripping the electrified tool, lock his jaws together, prevent him from calling out for help, and slowly interrupt his regular heartbeat until he is electrocuted.

Heavy Construction Equipment

Heavy mobile construction equipment, such as cranes, front-end loaders, dump trucks, bulldozers, and tow trucks, are required at every major collapse rescue operation. First, the vehicles surrounding the collapse area will be towed away in order to allow heavy equipment to be brought close to the collapse site. Bulldozers may be used during selected debris removal to shift rubble taken from the collapse pile by firefighters. The bulldozer will clear a sidewalk of rubble and pile it at a remote loading point. Here, a payloader will be required to load the nearby dump trucks, which must have space to park while awaiting loading. Road access in and out of the collapse area will be required. Finally, a crane and bucket will be moved into the collapse area for general debris removal.

A general rule of safety during a collapse rescue operation states that heavy mechanical equipment should not be used on the same collapse rubble pile where rescue workers are also digging with hand tools for buried victims (Figure 2.6). Either mechanical equipment, with one or two firefighters as guides, will remove the collapse rubble, or firefighters will remove the collapse rubble by hand. Operating a crane, payloader, or tow truck near firefighters engaged in hand-to-hand digging is extremely dangerous and should not be permitted. Just as, when fighting a fire, large outside master streams that can cause a burning building to collapse should not be used while firefighters are inside the building using interior hand lines, so should operation of heavy mechanical equipment on a collapse rubble pile be avoided while firefighters are working on the pile. Use one type of rubble removal at a time: either heavy mechanical equipment or hand digging, but not both simultaneously.

When a crane and bucket are used during general debris removal, the bucket picks up the rubble from the collapse pile and dumps it at the central loading point. At this point, firefighters must examine the debris for evidence of victims. After examination, payloaders fill up the dump trucks. The location where the collapse rubble is being dumped must be made known to the officer in command of the rescue operation.

Firefighters, except for guides, should be removed from the area of heavy equipment use. Heavy equipment employing wire cables to pull cars or large pieces of collapse rubble may snap and break, and a cable whipping through the air can decapitate a firefighter. Firefighters could be crushed beneath the wheels of a piece of heavy equipment, since such equipment will not be required to move in the

Figure 2.6. Heavy mechanical equipment such as cranes should not be used on the same collapse rubble pile where rescue workers are digging with hand tools.

normal traffic patterns. Furthermore, large pieces of wood and steel often drop from the clam bucket of a crane while it is moving material from the collapse rubble pile to the central loading point.

Because of the dangers presented by use of heavy construction equipment at a collapse site, a chief or company officer should be placed in charge of its operations. The officer should coordinate its use with other phases of the collapse rescue operation and ensure that this equipment is used in a safe manner.

Search of Below-Grade Areas

During a building collapse caused by a tornado, earthquake, or aerial bomb attack, people are advised to take refuge in the cellar. The cellar of a structure will often remain unaffected by collapse of the walls and floors above. The masonry foundation walls of a cellar may be thicker than the walls above, they cannot collapse outward because they are reinforced by the earth surrounding them, and the floor separating the cellar from the first floor is often reinforced with steel columns and girders so that it can support a heavier load than the upper floors.

A cellar must be searched for trapped people because this area creates the largest and strongest void in a collapsed building, and there is a distinct possibility that people may have taken refuge there. Access to a cellar can be accomplished from outside the building even when the entrance hallway has collapsed. Cellars are sometimes accessible from sidewalk trapdoors, from cellar stairs in the rear yard, or from ground-level window wells at the side.

Accessing the cellar may lead firefighters to an interior cellar stairway that, in turn, could lead upward into the interior voids of the collapse structure. Hours of digging down through a collapsed roof can be avoided by entering a cellar, then climbing up the cellar stairs into the voids of the collapse structure in order to free trapped victims below a fallen roof or top floor.

Firefighters must realize, however, that entering a cellar during a major structural collapse is an extremely dangerous operation. The below-grade atmosphere could well contain carbon monoxide from a fire above. The oxygen in the cellar could be depleted, replaced by a gas, such as carbon dioxide, that will not support life. A combustible gas such as propane or methane may have filled the cellar; even though such a gas may be too rich to explode, it could replace all the oxygen, asphyxiating a firefighter who entered the cellar without a self-contained breathing apparatus already in place. Toxic liquids from broken containers on the floor above might be leaking down into the cellar area.

Water from broken pipes, too, will accumulate in the cellar area. If the electricity supply to the collapsed building has not been shut off, a firefighter standing in water who touches an electric supply line could be electrocuted. If the structure contains a subcellar (a cellar below the cellar) and that is filled with water, a firefighter may not be able to see the subcellar stair entrance and could fall in. Submerged in the water beneath the cellar floor, the firefighter will have only a minute or two to relocate the subcellar stairs, or he will drown, trapped beneath the ceiling of the subcellar. When walking on a cellar floor covered with one or two inches of water, a firefighter should probe the floor surface beneath the water before taking any forward step.

Before searching a below-grade area, a firefighter should ensure that the electric supply to the collapse building has been shut off; use a search rope so rescuers can locate him if he were to become trapped;

be equipped with a self-contained breathing apparatus, a flashlight, and a portable radio; and remain in constant communication with a supervisor.

Falls, Cuts, and Abrasions

Most injuries that firefighters receive during a collapse rescue operation are caused by falling or stumbling while climbing the uneven surface of the collapse rubble pile. This danger is not as great as a secondary collapse or an explosion and fire; however, if a second collapse, explosion, or flash fire should occur while firefighters are working atop a collapse rubble pile and are ordered to retreat quickly back down off the pile, many injuries will occur because of falls and tripping.

During such an emergency evacuation of a collapse rubble pile, firefighters may break bones and suffer puncture wounds and deep lacerations from falling on the pile's irregular surface. That surface consists of chunks of masonry, wood timbers projecting up out of the debris, twisted pieces of steel and wire cable, broken sections of plaster walls and wood lath, pointed nails, boxes, and broken pieces of furniture.

Firefighters climbing to the aid of injured people cannot choose the safest path. The rescuing firefighter must look ahead, guess what piece of debris seems most stable, step on it, and then hope it supports his weight. At night, of course, movement becomes more hazardous. During a nighttime collapse rescue operation, the rubble pile should be well lit with floodlights and spotlights to reduce injuries due to falls and tripping.

To minimize the incidence of falls, firefighters should reduce the amount of walking they do on the uneven collapse rubble pile. Instead of walking up on the pile, picking up a piece of debris, and then walking off the pile, firefighters should be lined up in a pass-along row, sometimes called a "human chain" (Figure 2.7). Debris should be passed from firefighter to firefighter while each remains relatively stationary.

Firefighter protective clothing offers another protection against injury caused by falls or stumbles on the collapse rubble pile. The padding of insulated firefighting clothing may soften the impact of a fall. A torn or ripped turnout coat or bunker pants may result from a fall, but this damage would probably have caused a cut or puncture wound if no protective clothing had been worn.

Figure 2.7. *Victims should be passed from firefighter to firefighter, lined up in a pass-along row. This arrangement is sometimes called a human chain.*

During the most dangerous stages of a collapse rescue operation in particular—site survey, surface search of the collapse rubble pile, and void search—protective clothing must be worn. There is a great danger of secondary collapse, explosion, or fire. During selected debris removal, the officer in command should consider allowing firefighters to remove their protective clothing, with the exception of helmets and boots, during a long digging operation in order to prevent heat exhaustion.

Uncoordinated Collapse Rescue Operation

When rescue workers arrive from several different organizations—fire department, police department, construction companies, emergency medical services—there is a danger of uncoordinated, uncontrolled rubble removal. Uncoordinated rescue work can be deadly. Some rescue workers might inadvertently pull collapse rubble down on top of other rescuers. Some rescue workers might unknowingly cut structural members supporting other rescue workers. Some rescue workers using mechanical equipment might accidentally injure other rescue workers digging by hand.

When several different groups are engaged in collapse rescue work, there should be one person in charge to coordinate the collapse rescue personnel operating at the collapse rescue site (Figure 2.8).

Figure 2.8. *When rescue workers are sent by different agencies, there is a danger of uncoordinated, uncontrolled rubble removal.*

This coordinator must be designated and recognized by those in charge of all the agencies, and all personnel digging on the collapse pile must be informed of that designation. His primary responsibilities are the rescue of all victims buried in the collapse and the safety of all collapse rescue workers; he should direct the collapse rescue operation according to the standard collapse rescue plan.

Lessons to Be Learned

A building collapse rescue operation is one of the most dangerous jobs to which a firefighter can respond. The dangers present at a collapse have been described above. The following list offers some safe operating procedures.

1. Stretch a hose line to protect rescuers from any sudden explosion or flash fire.
2. Follow the standard collapse rescue plan:
   a. Site survey
   b. Collapse rubble pile surface search
   c. Void search
   d. Selected debris removal; tunnel/trench to specific locations
   e. General debris removal
3. Shut off all utilities—gas, electric, water—as soon as possible.
4. After site survey, surface search, and void search have been completed, temporarily withdraw all rescue workers from the pile. Increased safety precautions should be taken; the most productive life-saving stage of the rescue plan is over. Although digging for buried victims continues, the risk (danger to firefighters)/benefit (chance of rescuing victims) relationship has changed, against the rescuers. Continue selected debris removal after serious secondary collapse, fire, and explosion dangers have been eliminated, and use fewer firefighters. (See Figures 2.9, 2.10, 2.11, and 2.12.)
5. During night operations or at major structural collapse rescue work that will be of long duration, obtain portable lighting equipment.
6. When there is a danger of secondary collapse, use a transit (surveyor's device) to measure any movement of the structure. If a wall shifts more than a quarter of an inch during a collapse rescue operation, rescue workers should be removed

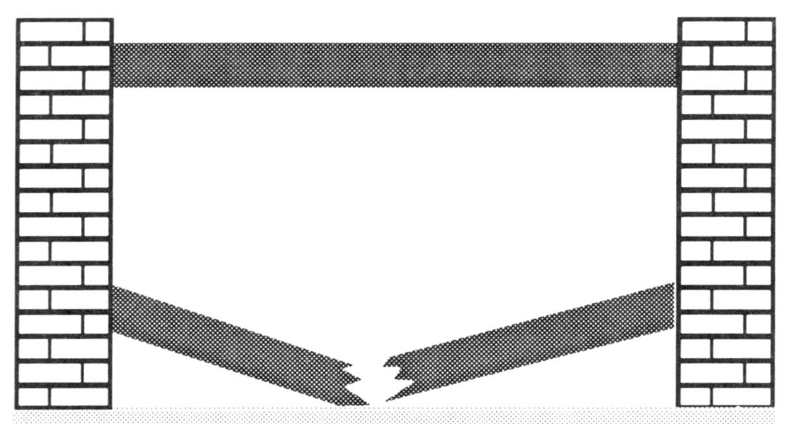

Secondary Collapse Danger: Bearing wall failure due to lateral load placed on bearing walls by broken floors. Floors may collapse further.

Victims found in voids near walls and at center of 'V'.

Cause of Collapse: a center floor overload.

Figure 2.9. *The secondary collapse danger of a V-shaped floor collapse is bearing wall failure, due to the lateral load placed on bearing walls by broken floor beam ends.*

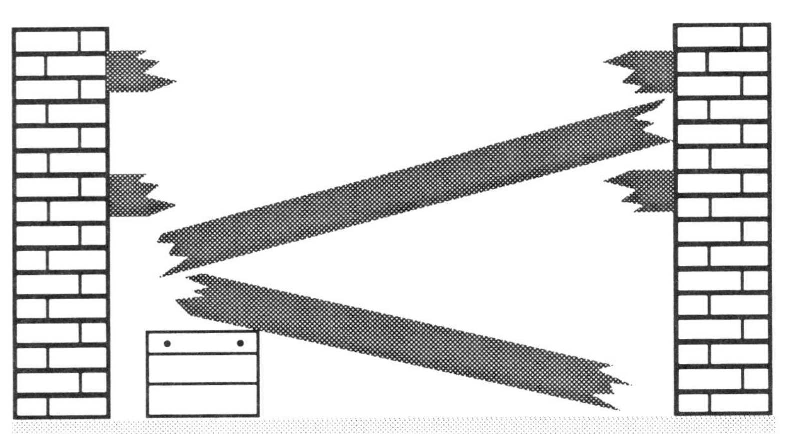

Secondary Collapse Danger: Collapse of free standing wall and pancaked floors collapsing further.

Victims found between floors in voids created by large pieces of furniture.

Cause of Collapse: Shock or impact of heavy falling weight.

Figure 2.10. *The secondary collapse danger of a pancake floor collapse is the collapse of a free-standing wall and further collapse of the pancaked floors.*

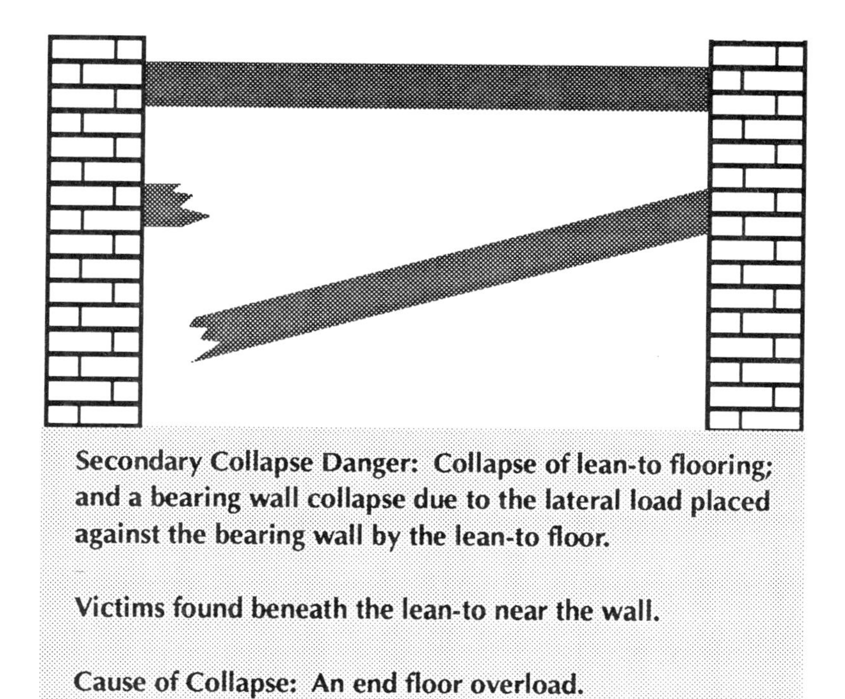

Figure 2.11. *The second collapse danger of a lean-to floor collapse is collapse of the leaning floor and also a bearing wall collapse, caused by the end of the leaning floor that is resting against the bearing wall.*

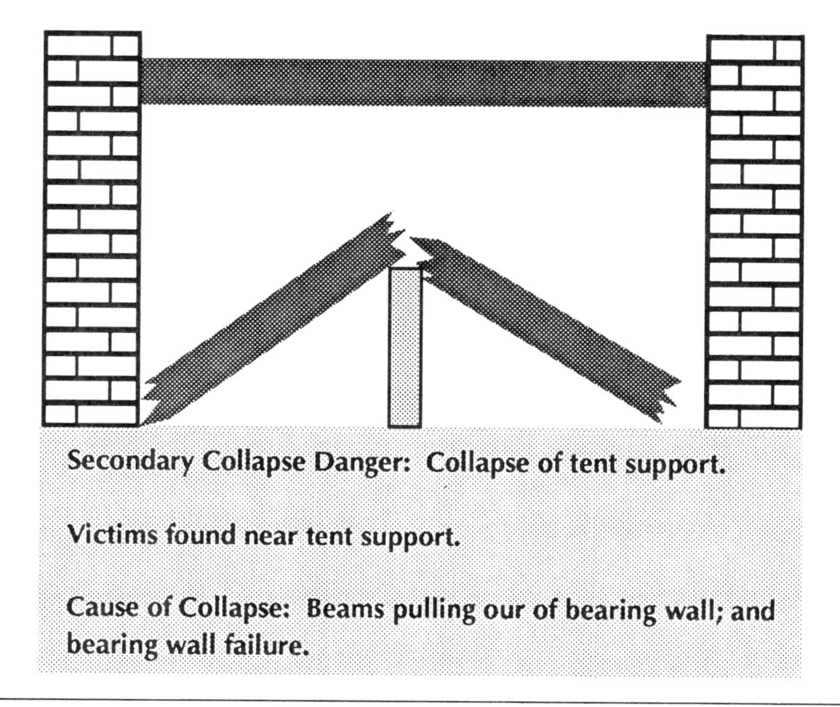

Figure 2.12. *The secondary collapse danger of a tent floor collapse is collapse of the tent support.*

Collapse Rescue Command System

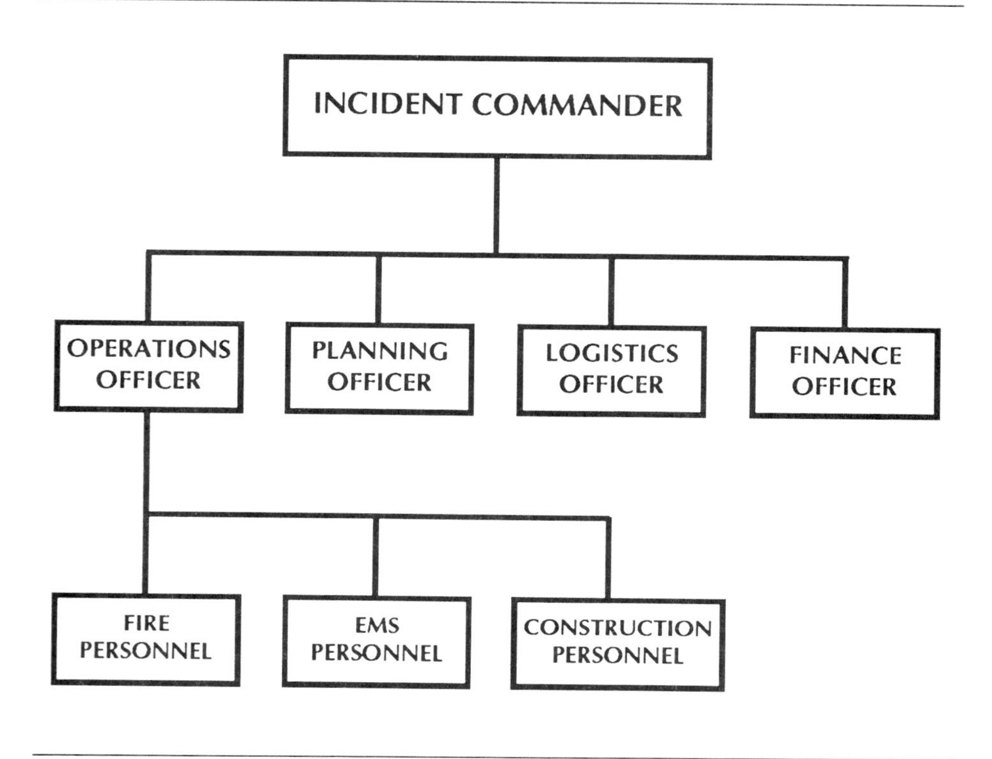

Figure 2.13. *A command system for use at a collapse rescue operation.*

from selected debris operation and the wall either shored or removed by a crane.

7. Set up a command system to coordinate the collapse rescue operation; use a command system modeled after the incident command system taught at the fire academy (Figure 2.13).
8. Designate a collapse rescue coordinator, who should be knowledgeable in collapse rescue operations. Notify all agencies that the coordinator is temporarily in charge of rescue operations.
9. Assign an officer to tracking record data on victims, such as name, location found, diagnosis, physician or medic name, facility to which transported, and by whom.
10. Do not use mechanical equipment to remove collapse rubble from the same area where hand digging is going on.
11. When there is danger of secondary collapse, do not pull down collapse structures in danger of collapse with wire

Figure 2.14. *A firefighter holds the arm of a woman survivor buried in a collapse for nine hours. Never give up hope of finding victims.*

cables. Shore them up, or, before selected debris removal begins, remove them with a crane after all rescue workers have been recalled from the collapse rubble pile.

12. Secure the collapse site before leaving the scene in order to prevent possible injury to civilians.
13. Photograph the collapse operations in order to obtain future visual training aids.
14. Never give up hope of finding victims. People have a good chance of survival if rescued within 24 hours; individuals have been buried in collapse rubble for as much as 11 days and lived (Figure 2.14).
15. Do not stop selected debris removal until all victims reported missing have been accounted for. During general debris removal, examine all rubble taken from the collapse for the remains of persons buried in the collapse but not reported as missing.

3

Responding and Returning Dangers

"GET OUT!" shouts the firefighter at the watch desk. "All companies respond to North Third and Water Streets! A reported house fire!"

He acknowledges receipt of the alarm, hits the button to raise the large firehouse doors, and hands the hard copy of the alarm location and address of the fire to the officer who is first to appear on the apparatus floor. Other firefighters converge on the apparatus floor from many different locations: Some arrive via the brass pole, some through doors of adjacent rooms, others from stairs to the second floor of the firehouse. They all head toward the pumper and the ladder truck parked in the center of the apparatus floor.

The apparatus engines start up. This sound adds urgency to the movements of the firefighters as they don their turnout gear. The firefighter at the watch desk quickly steps into his boots, jumps on to the back step of the pumper, and secures his helmet just before the pumper rolls out of the firehouse into the summer afternoon traffic. Even with siren wailing and red lights flashing, several blasts of the air horn are needed before the passing cars stop and allow the pumper to enter the flow of traffic.

The firefighter on the pumper's back step slips his left arm into the metal hand grip above the hose bed, tosses the turnout coat over his right shoulder, and slips his arm into the sleeve. He repeats the action for the other arm, all the while maintaining a hold on the hand grip. As the pumper enters the highway and picks up speed, the firefighter snaps the clips of the turnout coat and pulls his gloves from his pocket. Now fully geared up, he faces the front of the responding apparatus, firmly grasping the hand grips. He leans backward slightly.

The pumper approaches the corner of North Street, on to which it will turn right. The firefighter on the back step signals a right turn with his outstretched arm to warn drivers behind the pumper to slow down. The apparatus accelerates after the turn. The front wheels hit a bump in the road and the firefighter quickly bends his legs at the knees to absorb the vibrations of the back step.

The responding engine approaches the main intersection of town with siren and air horn sounding. The traffic light is in its favor. Vehicles approaching the intersection via other roads are partially hidden from view by the large trees that line the street. Through the tree branches, however, the firefighter on the back step sees a trailer truck heading toward the intersection at high speed.

The traffic light ahead of the fire apparatus turns yellow, and the chauffeur sounds the air horn harder and longer. The firefighter on the back of the pumper, realizing the chauffeur doesn't see the oncoming trailer truck, moves to the side of the back step, away from the approaching truck.

The apparatus chauffeur suddenly catches sight of the oncoming truck and hits the brakes. The trailer truck driver sounds several blasts of his air horn and hits his brakes. The pumper skids forward. The trailer truck jacknifes out of control.

The sickening sound of metal striking metal cuts through the afternoon air as the trailer truck rams the side of the fire apparatus. The impact of the collision spins the pumper around 90 degrees. The firefighter on the back step is lifted into the air. His hands are torn from the hand grips and he is hurled forward. Arms outstretched, he flies 50 feet and slams down on the pavement. His helmet and boots fly from his body upon impact. Streaks of blood trail the firefighter's body as it slides, face down, across the road. His head strikes one tire of a halted auto, snapping his neck. The lifeless body continues its slide under a parked car.

Defensive Responding

A firefighter makes many life-and-death decisions during a tour of duty, but the first and most important one is climbing aboard a fire truck that's been called to respond to a fire. Whether you have your turnout gear on or not, whether you choose to ride the back step or the side passenger compartment, whether you snap on the seat belt or the restraining belt on the back step—these are some of the first life-and-death decisions a firefighter makes during a tour of duty.

Many firefighters believe that the chauffeur is responsible for their safety during response to alarms and return to station. This belief is wrong. A close look at reports on firefighter fatalities and injuries during these situations reveals that many of the tragedies are caused or could have been prevented by the firefighter who was killed or injured, not by the chauffeur of the apparatus.

In the past several years, the fire service has been affected by so-called "right-to-know laws" passed by the federal government. They tell us, among other things, that we have a right to know about hazardous materials stored in our communities and about toxic emissions given off by diesel exhaust fumes of the fire apparatus in quarters. Let's add to the list: We have the right to know that *we* are responsible for our own safety and survival when responding and returning—the driver or the officer is not. We are not passengers on the fire apparatus. The driver and the officer do *not* say, "Leave the driving to us."

Just as every driver of a fire apparatus should have a "defensive driving" attitude, so should every firefighter who climbs aboard maintain a "defensive responding" attitude. What is defensive responding? It's knowing that certain riding positions on a fire truck are more dangerous than others and using the safest one available; it's knowing the safest side of the fire truck to climb aboard when responding; it's knowing how dangerous roadway intersections are to a responding fire apparatus; it's knowing how to assist the driver of the fire truck through a narrow roadway or past double-parked cars; it's knowing how to operate safely at a fire on a high-speed highway. Defensive responding means realization that responding to and returning from alarms is just as dangerous as the hazards faced on the fireground itself. It's knowing that the entire company, not just the driver and the officer, are responsible for the safe arrival of the apparatus at the scene of the fire and its safe return to station. Safe responding, like safe firefighting, is a team effort.

Riding Positions on a Fire Apparatus

The decision to ride either in one of the fire apparatus enclosed passenger compartments (when available) or on the back step is critical. A collision with another vehicle or a sharp turn could cause a firefighter on the back step to lose his grip, in which case he will most likely be killed or permanently injured by the fall to the roadway or will be run over by another vehicle. If the firefighter secures himself to the apparatus back step by a restraining device, a spinal cord injury could occur during a collision. When a firefighter has a choice (many fire apparatus offer only the back step and do not possess a passenger seat compartment for firefighters) and yet climbs aboard the back step of a pumper or the side step of a ladder truck during a response, a bad decision has been made.

There are four riding positions on the modern fire apparatus. Each provides a different degree of safety during an accident, and firefighters should be aware of these differences.

Riding position on apparatus	*Degree of safety during collision*
1. Enclosed cab, secured with belt	Most ↑
2. Open cab, secured with belt	
3. Back step or side step, secured with restraint device	
4. Back step or side step, without restraint device	↓ Least

Some fire departments prohibit firefighters from riding on the back step or the side step of a fire truck (Figure 3.1). These departments already have or are planning to purchase vehicles with enclosed seating for all responding firefighters. In any event, company or department policy should mandate riding positions on all existing apparatus based on degree of safety in case of an accident. If the choice is still yours, ask yourself, "Is this the safest position I can choose?"

Climbing On and Off a Moving Fire Apparatus

The safest time to climb on to and step off a fire truck is when it's stationary. At the fire station, responding firefighters should quickly assemble on the apparatus floor near the vehicle and don all protective clothing. Then they should mount the apparatus—while it is stationary. The officer, when assured that everyone is aboard, gives the signal to the driver to respond.

Some departments that do not have traffic control signals activated from quarters use another boarding procedure, which requires that firefighters assemble on the apparatus floor, don all protective clothing, and assemble outside to halt traffic. The apparatus driver moves the vehicle to the middle of the roadway after traffic has been

Safety of the Riding Positions on a Modern Fire Apparatus

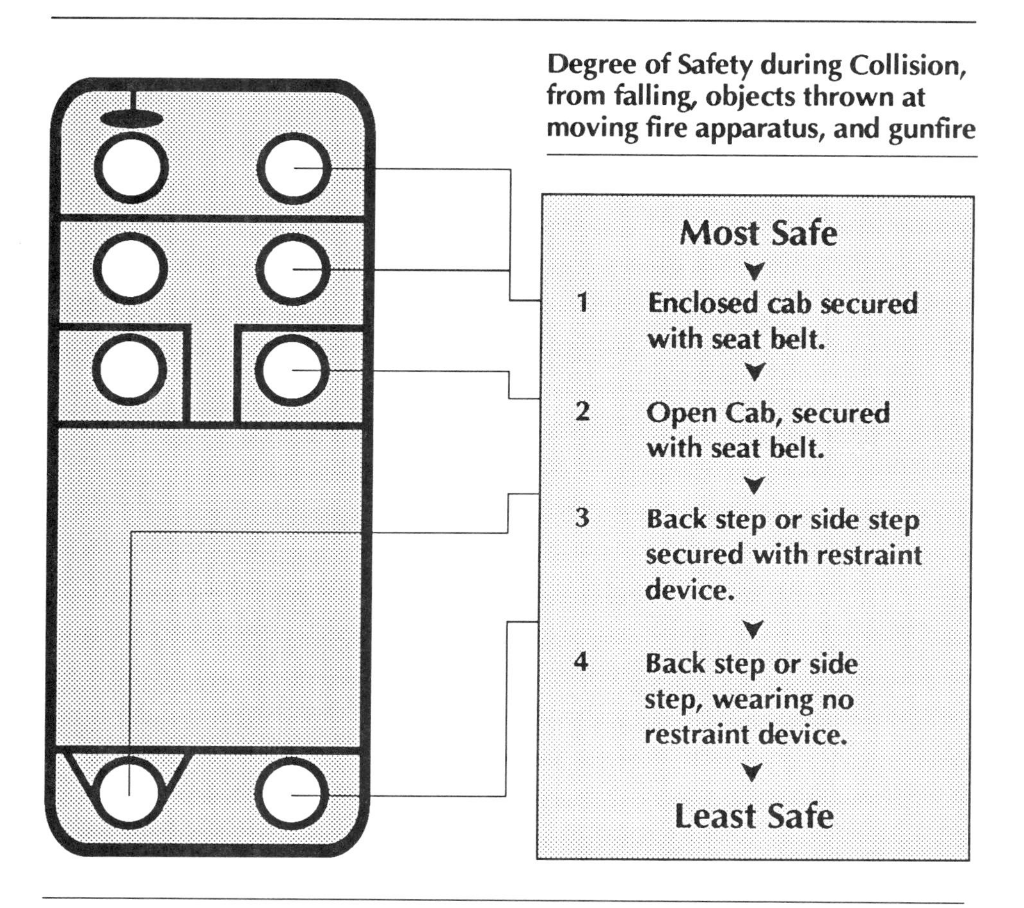

Figure 3.1. *Whether a firefighter chooses to ride the tail board or the side passenger compartment is a life-or-death decision.*

stopped. At this point, all firefighters climb aboard. The signal is given to the officer when everyone is safely on the apparatus. Then he gives the order to respond. In either case, firefighters should only board the vehicle when it is stationary.

During the course of a firefighter's career, however, he may have to jump on to or step off a slow-moving fire apparatus. Firefighters should know that this practice can be deadly even when the vehicle is moving slowly. They should know that firefighters have been crushed to death beneath the wheels of moving fire apparatus after failing in their attempts to jump aboard. Deadly head injuries have been suffered by firefighters who step from fire apparatus before the vehicles have come to complete stops: The firefighter steps off the moving vehicle with one foot, is propelled forward, and cannot get the other foot down on the ground fast enough to break into a run. He pitches forward, striking his head on the ground.

Defensive responding requires that a firefighter respond quickly to the apparatus floor when an alarm is received in the firehouse, so that he has adequate time to don turnout gear and get on the apparatus before it begins to move. After the fire has been extinguished, the defensive responder is always alert to the officer's signal to board. Chances of injury are great when the firefighter, having missed the signal, must catch up to the moving vehicle. If he is only able to run fast enough to get one foot up on the back step but not the other, he could fall forward and strike his head on the edge of the moving apparatus.

Climbing Aboard a Turning Fire Apparatus

Jumping on to or off moving apparatus is dangerous enough; climbing aboard a turning apparatus is even more so, and attempting that on the side that's "into the turn" is more dangerous still. If the firefighter misses the step of the moving fire apparatus and falls to the roadway, chances are great that the rear wheels will run over him before the driver can stop the vehicle (Figure 3.2). A firefighter who falls in an attempt to board an apparatus from the outside of its turn will be moving away from the direction of the vehicle and, although he still risks injury, at least the rear wheels are not likely to pass over him.

The officer in command of the fire apparatus is usually held responsible by department personnel investigating the types of accidents mentioned above. The officer must always be assured that all firefighters are properly and safely boarded before giving the order to proceed. In most instances, however, the firefighter who falls and is injured or killed must also bear some responsibility.

Operating at the Scene of a Highway Accident or Fire

Firefighters operating at a motor vehicle fire or extrication on a highway are in the greatest danger not from flames or leaking gasoline but from oncoming, speeding trucks and cars (Figure 3.3). Each year, firefighters crossing highways during emergency operations are struck by passing vehicles and killed. Firefighters operating hose lines are

Danger Side of a Moving Apparatus

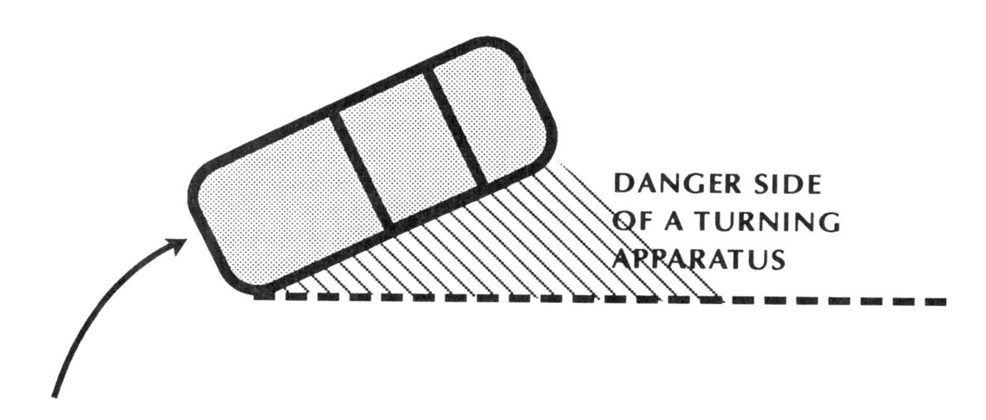

Climbing aboard a turning fire apparatus is extremely dangerous.

Climbing aboard the side of the apparatus that is "into the turn" is more dangerous still. If the firefighter misses the step of the moving fire apparatus and falls to the roadway, he may be run over by the rear wheels before the driver can stop the moving apparatus.

Firefighters should board apparatus when it is not moving.

Figure 3.2. *Climbing aboard a moving fire apparatus on the side that's "into the turn" is extremely dangerous. If you miss the truck and fall, the rear tires will run over you.*

Figure 3.3. *The greatest danger to firefighters extinguishing a car fire on a highway is not from the flames or leaking gasoline but from oncoming speeding cars and trucks.*

crushed to death when speeding vehicles rear-end a parked fire truck.

When a fire company arrives on the scene of a highway emergency and there are no police at hand to control traffic, firefighters themselves must first control the oncoming vehicles before safely turning their attention to the emergency. The fire apparatus should be parked on the highway shoulder, if possible. Warning signals or flares should be placed in the roadway to alert oncoming vehicles. A firefighter directed to stop highway traffic must never turn his back on oncoming vehicles, nor should he believe that the oncoming vehicles will stop for red warning lights, flashlights, or flags, or because he's dressed in firefighting clothing. He must always face oncoming traffic and be prepared to jump out of the way at a moment's notice.

Roadway warning devices must be properly and prominently displayed. The firefighter must judge the distance required to stop a vehicle traveling 60 miles per hour. He must also consider the effects that a downgrade, a curve in the road, darkness, and various weather conditions have on the stopping distance. Warning devices on a high-speed highway should be placed at least 350 feet from the fire apparatus and positioned so that they are visible to an oncoming motorist for at least a further 350 feet before that. This placement will give the driver 700 feet to stop a vehicle once he perceives the danger. A curve or upgrade requires that the warning signal be placed farther than 350 feet away from the fire apparatus.

To determine a vehicle's total stopping distance, three factors must be considered:

1. Perception time: The time it requires for the driver to realize a danger exists on the roadway. This time is estimated at ¾ of a second.
2. Reaction time: The time needed by a driver to move a foot from the gas pedal to the brake. This interval is also estimated at ¾ of a second.
3. Braking distance: The distance a vehicle will move after braking.

If a flare or warning device is placed in the roadway 350 feet before the emergency scene and it is visible 350 feet beyond that point, a truck moving at 60 mph will stop just before the warning device. The additional 350 feet between the warning device and the apparatus will provide a safe distance within which to operate and will also provide a safety factor against vehicles with defective brakes or against motorists who are driving while impaired and cannot see the emergency scene (Figure 3.4).

Placement of Warning Signs on Highways

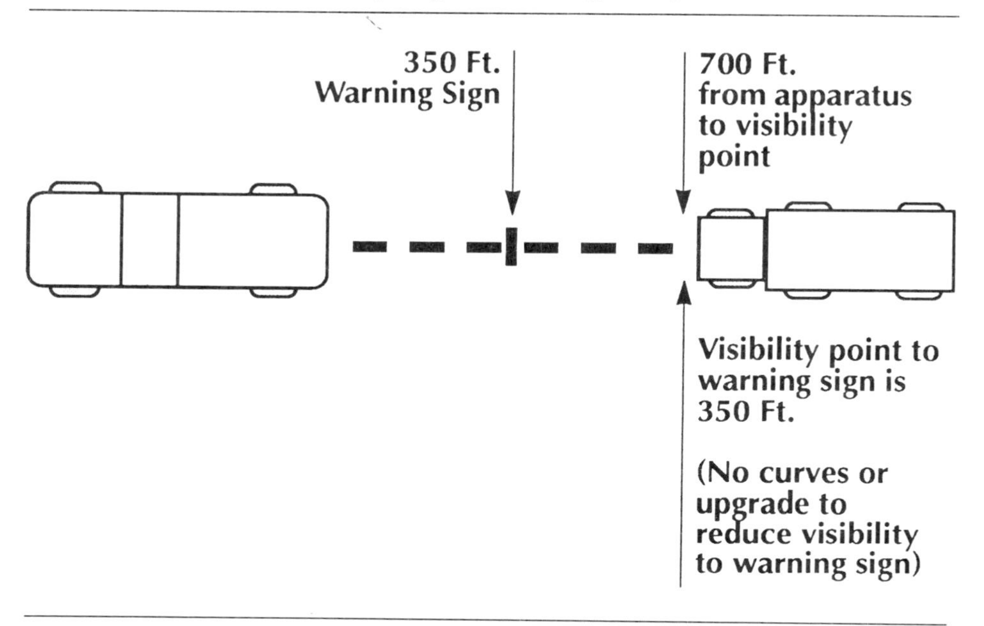

Figure 3.4. *Warning devices should be placed at least 350 feet behind the fire apparatus and positioned so they are visible to oncoming traffic for at least 350 feet beyond that. This placement gives the driver 700 feet in which to stop a vehicle.*

Intersections

Highway/roadway intersections are extremely dangerous places for responding firefighters and civilian motorists alike. The intersection accident has proven to be the most deadly to all concerned.

When the responding apparatus approaches an intersection against a traffic signal—even with red lights and sirens displayed, as required by law—the apparatus chauffeur must be certain that the right of way has been yielded by all oncoming motorists before proceeding into the crossroad. The defensive fire department driver downshifts or slows down, gently braking, upon approaching an intersection. In that manner, if traffic fails to yield the right of way to the apparatus, the chauffeur will be able to make a complete stop before reaching the intersection (Figure 3.5).

The defensive fire department driver slowly proceeds into the intersection and passes through it only after he is certain that all other motorists have seen the fire apparatus, are slowing down, and are capable of coming to a complete stop. Some fire departments require that drivers bring the apparatus to a temporary stop at every intersection when responding, for the benefit of both firefighters and civilians.

Figure 3.5. *At a roadway intersection, the "defensive driver" slowly proceeds into the intersection and passes through it only after all other motorists have seen the fire apparatus, are slowing down, and are capable of coming to a complete stop.*

Squeeze-Through

The "squeeze-through" accident is the most common collision involving fire apparatus. It occurs when the apparatus must squeeze through a narrow space—a street with cars parked on both sides or even double-parked, for instance. If the size of the restricted roadway is misjudged or the vehicle is not maneuvered through the space at the correct angle, the fire truck will strike one or more parked cars.

There are usually no injuries at a squeeze-through accident; the damage is often minor. The ramifications of such an accident, however, can be of major importance to fire department operations and life safety at the scene of a fire: The law requires that the fire department remain at the scene of the accident. If the accident is minor, the officer in charge of the company, then, must do one of two things: either order the entire company and apparatus to remain at the scene of the accident for the official police investigation or order that one firefighter stay behind while the rest of the company continue the response, to return later after the response has concluded. In the first instance, the community loses the services of the fire company at the fire scene. In the second, the company arrives at the fire minus one firefighter. For a three-member company, the second alternative means that they will be seriously undermanned, and that could mean the difference between life or death at a serious fire.

Responding firefighters who encounter a restricted roadway should take several defensive actions to assist the driver and to prevent a collision. They should get off the apparatus and request the driver of the double-parked car or truck to move. If the driver is not in the vehicle and there is enough room to pass, the firefighters should guide the driver through the restricted area of roadway. The company officer can also assist the driver in this situation; he, not the driver, is responsible for determining whether there is sufficient space for the apparatus to proceed. If there is, the officer should shut off audible warning devices so the driver can concentrate on the difficult maneuver. Company firefighters should be positioned to guide all sides of the apparatus. If the officer decides that there is not sufficient space for the apparatus to proceed, he must order the driver to back out of the street, guided by the firefighters. The communications center should be notified of the delay in response.

This sometimes time-consuming procedure may appear to contradict everything we are taught about quick response and receiving the right of way by sounding the apparatus sirens. Yet the service a fire company provides for a community is based on dependability, not speed. Arriving at the scene of an emergency when called—not how long it takes us—is the number-one priority of a fire department. A momentary delay is better than never arriving at the scene because you've become involved in a squeeze-through accident.

The Fire Apparatus in Reverse

In the aftermath of squeeze-through accidents, many review boards have assigned responsibility to the officer if the collision has involved the right side of the apparatus and to the driver if the collision has involved the left side. Accidents that occur when the apparatus is operating in reverse gear are most often attributed to the fault of firefighters other than the driver.

Firefighters should never be riding on an apparatus that is operating in reverse; they should be off the rig and in the roadway, assisting the driver. The dangers associated with such a common action must never be underestimated. Anxious motorists have attempted to pass apparatus that have stopped momentarily to reverse gears. Firefighters stepping off the fire truck have been struck by such oncoming automobiles and killed.

Before the returning apparatus backs into quarters, all traffic must be stopped by firefighters. Defensive responding requires that they face oncoming traffic and use flashlights at night; they must be ready to jump out of the path of an auto driven by a drunken or reckless driver (Figure 3.6). After traffic has been stopped, one of the firefighters should take a position near the firehouse entrance doors, on the officer's side of the apparatus, to help guide the driver and warn him should it appear that the vehicle will strike an object. This

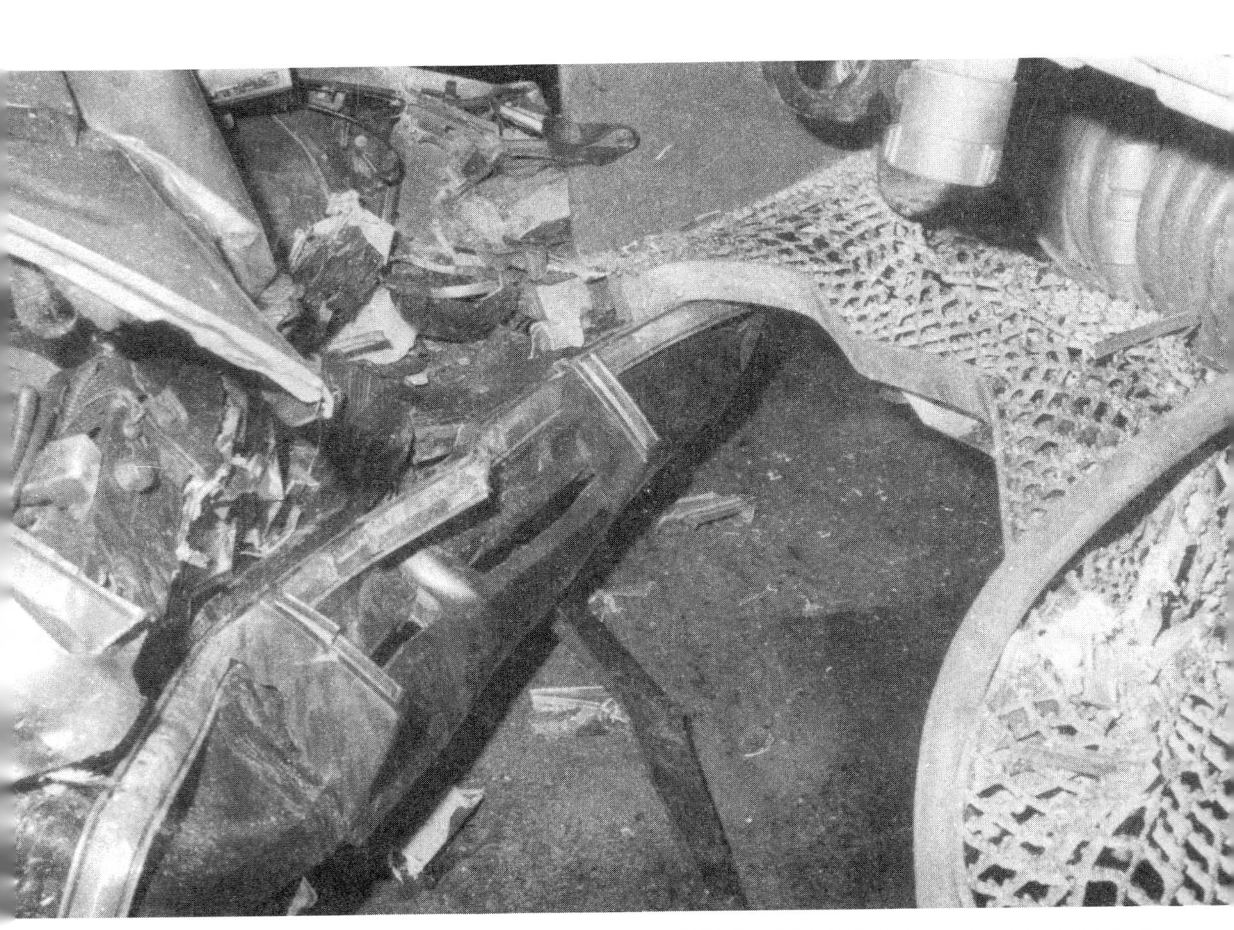

Figure 3.6. *Firefighters riding the tail board of a fire apparatus, such as this one, have been struck and killed by oncoming vehicles which failed to see warning lights.*

firefighter is also responsible for preventing impatient pedestrians from dashing into the path of the fire truck. Only when the fire apparatus is completely off the roadway and in the firehouse should the order be given to allow traffic to proceed. Firefighters in the roadway should face waiting traffic during the entire backing-up operation. They must be prepared for a surprise move by an over-anxious motorist in a line of traffic, who sees that the apparatus has pulled into quarters but is unaware that firefighters are still in the road.

Lessons to Be Learned

The hazards of responding to and returning from alarms are increasing. They are too great to place the responsibility for safety solely on the firefighter operating the vehicle. Defensive responding requires a team effort by all members on the apparatus, for the safety of all is at stake.

4

Searching for the Fire Location Dangers

"All hands come to the housewatch," the firefighter calls on the firehouse public address system. "We have a fire across the street." The captain of the engine company runs out of the housewatch area and asks, "What's up?" The firefighter on housewatch points to a man standing in front of the open firehouse door and says, "Captain, this man just reported a fire across the street."

"Yes, please hurry; my apartment is on fire. The building is number Two Main Street, apartment sixteen."

The captain turns to the firefighter on housewatch and says, "Okay, let's go. Turn out the company." The pumper is driven out of the firehouse into the hot summer afternoon sunlight. Looking up through the front window of the moving fire engine, the officer can see smoke pushing out around the top-floor window frame of a large, 30- by 50-foot, attached row house of ordinary construction. The pumper stops at a hydrant, and the officer jumps out of the truck, adjusts the shoulder straps of his SCBA, and heads for the front stoop leading up to the doorway of number Two Main Street.

"Start the stretch," he calls to the firefighters assembled at the rear of the apparatus. The captain quickly runs up the steps to the building, enters the cool, dark, tile hallway and starts up the interior stairs to the third floor, where the fire has been reported. People carrying small children, television sets, dogs, and cats are coming down the stairs. "The fire is on the top floor," they call out as they pass the officer climbing the stairs, two steps at a time.

Arriving at the top-floor landing, the officer sees smoke puffing out of a partially opened door with the number 16 painted on it. He crouches down, pushes the door further open, smells the familiar odor of burning paint, and looks down a long hallway. Smoke has banked down halfway between the ceiling and the floor. The bottom portion of a window is visible beneath the smoke at the front of the apartment. The officer dons his facepiece, switches on his handlight, and begins to crawl down the hallway beneath the smoke. If he can make a quick search, locate the exact room on fire, and isolate it by closing a door before it flashes over, that action can make a difference in the operation. Success in these operations could mean a quick, one-room "knockdown," as opposed to a tough hose line advance down the hall of a fully involved apartment.

As the captain moves farther into the apartment, he momentarily raises a gloved hand up into the smoke. He feels more heat in the smoke than he had expected. Brushing against the right side of the solid hallway wall, he crawls forward and looks to his left and up at the ceiling for any sign of a red glow. The black smoke continues to descend in the apartment; the beam of the handlight is barely visible. "I'd better find the room fast," thinks he. Bumping into a sofa and moving around it, he regains contact with the wall on his right. He can no longer see the bottom of the front window beneath the smoke. The smoke has banked down to the floor. A red glow is visible to his left. Quickly moving across the open room, the officer reaches up, feels for the doorknob, finds it, pulls it—and the doorknob comes off in his hand. Next, he reaches outside the room with a gloved hand, grabs the bottom edge of the door, and attempts to pull it closed. The door is stuck. It closes slowly.

"I'd better get out." Just then a large section of hot plaster ceiling crashes down on the officer's helmet. He is dazed. Hot pieces of plaster begin to burn his neck and back after rolling down his upturned collar. He reaches back to brush the hot plaster off his neck and collar. *Boom!* A flash of fire shoots out the door into the officer's facepiece. Caught off balance, he falls backward. The heat banking down from the ceiling becomes unbearable. The window! Where is the window?

Crawling along the floor on elbows and knees through the pitch-black smoke, the officer bangs into a wall with his head. Reaching up with one hand, he feels for a window and finds only a solid wall. He thinks to himself, "Calm down! I can't lose my head now! Let's see. I came into this room searching counterclockwise. I should find my way out going in a clockwise direction—to my right." Hugging the wall with his body to avoid the heat above, he crawls ahead several feet, pushes over a small table that is in his way, then knocks over a floor lamp in order to stay close to the wall. Reaching up again with his left hand, he feels a raised molding and a depression in the wall—a window.

Pulling himself up on the windowsill with his left elbow, he forcefully swings his gloved right hand, holding the handlight, through the lower glass windowpane. Quickly knocking out the remaining jagged edges of glass around the frame, he climbs up, leans out of the smoke-filled window, pulls off his helmet and facepiece, hangs his head below the smoke coming out the window, and feels the warm summer air on his sweaty face.

Looking down 30 feet to the street level, he can see the pump operator hurrying around the engine with the female end of the hose line about to be connected to the discharge outlet. "The hose line'd better knock this fire down quickly, or I am going to have to choose between burning or jumping," he thinks. Raising himself up, he pushes his shoulder and chest out of the window. The heat mixed with smoke flows out of the window above his head; he feels it on his neck.

"Hey, Joe," he calls to the pump operator working in the street below. The pump operator does not look up but continues checking the pump gauges and supplying water volume and pressure to the hose line. The captain's call is drowned out by the pumping engine.

"Joe!" he calls again. The firefighter looks up, sees the trapped officer and cries, "Stay there, captain; I'll get a ladder up to you." The smoke billowing out the window becomes hotter. The officer knocks the remaining pieces of glass from the window frame and swings his left leg out the window. Now straddling the sill, his head positioned low to avoid the heat and smoke coming out the window, he thinks, "Where is that hose line? It should be in here by now."

Looking down into the street, his question is, unfortunately, answered. A spray of water shoots up from the hose 20 feet into the air. A burst length! The pump operator is struggling with a ground ladder, pulling one end of the ladder off the side of the pumper, laying it on the sidewalk, running to the other end of the ladder, and removing the lock fasteners.

Just then the wind shifts; smoke and heat engulf the captain caught at the window. Although he closes his eyes tightly, the smoke still irritates. He begins to choke on the acrid, black smoke. He shifts farther out the window to reach fresh air. His left hand searches for the dangling facepiece, holding on to the sill with his right hand. The wind shifts again. He is in the clear. The smoke flowing out the window is now above his head.

He can feel his right shoulder and leg begin to pain from the heat. He thinks, "I don't know how much longer I can hold on." Looking down to the street 30 feet below, he sees a metal awning over a ground-floor window. He wonders if he could manage to hit it as he drops. "It could break my fall," he thinks to himself. Directly below his window is a spiked fence enclosing an area containing garbage cans. "If I have to jump and do survive, I will probably never be a firefighter again."

Figure 4.1. *When rooms become superheated before flashover, firefighters are forced to crash through the nearest window. Unfortunately, there is sometimes neither a ladder nor a fire escape on the other side.*

The siren of a fire truck interrupts his train of thought. A ladder truck pulls into the street. The pump operator drops the ground ladder, runs over to the firefighter positioning the truck, and points up to the smoke- and flame-engulfed window, with the captain straddling the sill. The top windowpane, above the captain, cracks in the heat, and pieces of glass fall on the officer. He ducks down to avoid them. They crash into the street; flames now begin to lap out the top of the window (Figure 4.1).

"Hold on, captain!" the pump operator shouts to him, as the metal aerial ladder is raised slowly out of the truck bed. The pain in the officer's right shoulder and neck have become unbearable. He starts to climb all the way out of the window. With both legs out, his knees and the toes of his boots pressed against the outside wall of the building, he shifts his upper body around. His left arm is placed inside, holding on to the sill. His right arm is out of the window, and he sees the right sleeve of his turnout coat charred and smoldering. "This is it. I can't hang on any longer!"

Suddenly, something touches the bottom of his boots. Looking down, he sees the tip of the aerial ladder. Quickly stepping on the narrow top rungs, he shifts his weight from the window sill to the ladder tip; the ladder sways into the building and bounces slightly, but it feels as solid as the Rock of Gibraltar to the officer. Two arms grab him from behind. "Captain, are you all right?" asks the firefighter on the ladder in back of him. "I am a helluva lot better than I was one second ago, thanks to you," replies the officer.

There are three flash phenomena that occur in a superheated, smoke-filled room, and all of them can trap and kill firefighters: rollover, flashover, and flameover. A firefighter should know the differences and characteristics of these three bursts of flame that can occur suddenly in a smoke- and heat-filled room.

Rollover

Rollover is a sudden, sporadic flash of flame mixed with smoke appearing at the upper ceiling level just before flashover occurs. It is a warning for firefighters to withdraw from the fire room. Fire protection engineers define rollover as flashes of combustible gases, released during the growth stage of a fire, that have mixed with air and are entering their flammable range. These intermittent flames are first noted just outside the doorway to a fire room. They are warnings of a more deadly event about to happen—flashover. Rollover can, however,

Figure 4.2. ABOVE TOP: *Flashover, defined as full room involvement in flame, has taken place when a smoke- and heat-filled room suddenly bursts into flame.*

Figure 4.3. BOTTOM LEFT: *Flashover signals the death of any victim or firefighter trapped in the room. Temperatures reach 1,000 to 1,500 degrees F. when flashover occurs.*

Figure 4.4. BOTTOM RIGHT: *Flashover signals the beginning of a structural collapse danger. After flashover occurs, the flames begin to attack the structure.*

trap firefighters in a cellar fire. Flashes of flame mixing with smoke, rising up out of a cellar entrance, may prevent firefighters from climbing back up a cellar stairway. Whenever firefighters enter a below-grade area, a firefighter should be assigned to remain at the top of the stairway in order to warn those below to withdraw if heat build-up or rollover flaming occurs.

Flashover

Flashover is the explosion of a smoke-filled room into flame; it takes place after rollover. Flashover signals the end of any effective search or rescue operation; it signals the death of any trapped victim or firefighter inside the blazing room; it signals the end of portable extinguisher use as a means of fire control—a hose line is now required to contain the blaze; it signals the beginning of a structural collapse danger; it signals the end of the growth stage of the fire (Figures 4.2, 4.3, and 4.4).

Fire protection engineers define flashover as the result of thermal radiation feedback (re-radiation) from ceiling and upper walls that have been heated by a fire in a room. During a fire, radiation feedback gradually heats the smoke, the fire gases, and the contents of the room. When all combustibles in the space have become heated to their ignition temperatures, sudden, simultaneous ignition or flashover of the entire room occurs (Figure 4.5).

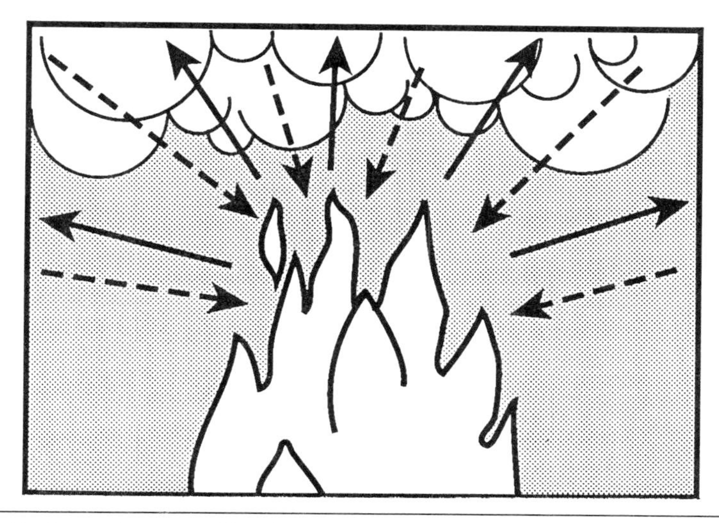

Figure 4.5 *Flashover is caused by thermal radiation feedback. Heat radiates back into the smoke-filled room from the upper walls and ceilings. When the combustible gases and content have been raised to their ignition temperatures by the radiated heat, flashover occurs.*

Time-Temperature Course of a Fire

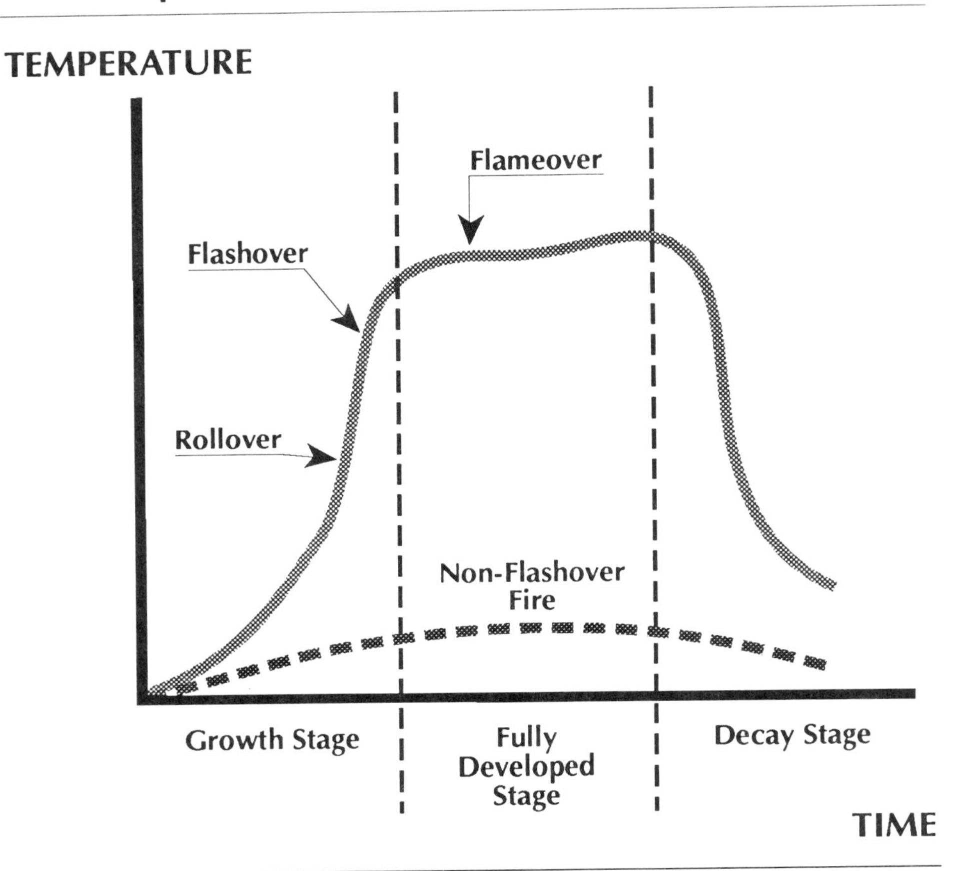

Figure 4.6. *There are three stages of a fire. Flashover signals the end of the growth stage. After flashover, the fire is in the fully developed stage.*

A fire can be divided into three periods: a growth period, a fully developed period, and a decay period. Flashover takes place at the end of the growth stage, just before the fully developed stage. It can be identified on the time/temperature curve as the almost vertical upward line of the graph in the growth stage of fire development.

Flameover

Flameover, the third flash phenomenon encountered by firefighters, is completely different from either rollover or flashover. Flameover usually occurs after flashover; defined as rapid flame spread over one or more surfaces during a fire, it is caused by the sudden ignition of combustible vapors that are produced from a heated surface. Ceilings, wood-panelled walls, the tops of school desks, theater scenery, and nightclub decorations that have an applied surface coating of polyurethane or other combustible veneers can exhibit rapid flameover when heated.

Once a room flashes over (Figure 4.6) and flames begin to spread out of the original fire area into adjoining spaces, flameover is a danger. Firefighters advancing hose lines down hallways or corridors to a blazing apartment have been encircled by flameover. Rapid flame spread over the heated combustible surface of the corridor walls and ceilings quickly spreads fire behind forward-moving firefighters operating the nozzle at the door of a burning apartment. Also, firefighters battling school classroom fires in coat closets have been burned by the rapid, simultaneous ignition of the tops of plastic-coated school desks—an example of flameover.

Flashover, the most dangerous of the three kinds of flash phenomena, is a particular danger to first-due firefighters, who make the initial entry into the burning structure and attempt to locate the point of fire origin. Frequently, they are fire officers sizing up the interior of the burning house or apartment to determine the life hazard and the best area for hose placement.

No one can predict when a flashover will occur. At some fires, the room may become a sudden inferno the moment you enter it, or the fire may enable you to search for long periods before flashover occurs. General statements describing time periods for flashover are intended to inform the public of the danger of rapid fire spread, but these statements are useless to firefighters as as safety guide; unlike the person who accidentally starts a fire, we do not know how long the fire has been burning.

The best safety measure a firefighter can take to determine the danger of flashover is to assess the heat build-up in the smoke-filled room. Very little heat mixed with the smoke indicates that flashover is unlikely. Very hot smoke, which forces a firefighter to crouch down to enter a fire area, indicates a danger of flashover. Experienced firefighters entering a smoke-filled doorway in a fire area should make mental note of the amount of heat they encounter. The lower they are forced to crouch below heat and smoke banked down under the ceiling, the greater the heat level and the chance of flashover. When crawling around a large, smoke-filled area searching for the fire's point of origin or for trapped victims, firefighters must crouch down below the smoke level for visibility; however, they should periodically raise a hand above their heads to check for a build-up of heat in the smoke, which indicates an increased danger of flashover.

Figure 4.7. Firefighters reported killed by flashover are often first disoriented by smoke, lose their sense of direction in smoke, their masks run out of air, and then they are killed by flashover.

Many firefighters who are reportedly killed by flashover are not in fact caught by a sudden blast of flame. First, they are lost in smoke. They become disoriented, lose their sense of direction, wander around in circles searching for the way they entered, their SCBA air runs out, and they are then burned to death in a flashover (Figure 4.7).

To reduce the danger of encountering flashover, a firefighter should be taught how to search a smoke-filled fire area without becoming disoriented and lost in dense, black smoke. When searching a small room in a private house, a firefighter should move consistently in a clockwise or counterclockwise direction with the entrance door of the room as the starting and finishing point. Keeping in contact with the wall may require the firefighter to climb over or move small pieces of furniture out of the way. Upon entering a long, smoke-filled corridor, a hand should be kept in contact with the wall; when leaving, the opposite hand should be in contact with the same wall.

When firefighters search large areas or mazelike cubicles, a search line should be used. One hundred feet of small-diameter rope with a metal snap clip at one end can be played out in a large, smoke-filled area as the firefighter penetrates deeper into the smoke. If the metal

Figure 4.8. *A firefighter can prevent the build-up of heat necessary for flashover by* venting*—preventing the build-up of heat in the room—or by* not venting*—starving the fire.*

snap clip is secured to an immovable object at the door or other entrance, the rope can be used as a guide to find the exit if smoke conditions worsen. A search line is necessary to search a smoke-filled floor in high-rise buildings safely. Even if smoke conditions on the floor are not severe, this condition could change speedily. The chances of rapidly finding your way out of a large floor area or through a maze of cubicles when smoke quickly banks down from a ceiling space or billows out of a large storeroom on fire are slight.

A firefighter may be reported as killed or seriously injured by a flashover, but actually the smoke-reduced visibility, causing disorientation, and the firefighter becoming lost in smoke precipitate such a death by flashover.

Ventilation and Flashover

It's important to prevent or delay flashover whenever possible. Since the build-up of heat from the fire is the key factor causing flashover, if you can prevent the build-up of heat in a burning room you can stop or delay flashover. The best way to achieve this is to get water on the flames and extinguish the fire quickly. The next best method is to prevent the build-up of heat by venting or, paradoxically, by not venting.

When you vent smoke and heat from a burning room by opening windows, doors, and roof skylight, that action can delay the build-up of heat necessary for flashover. If the room in question has a skylight, vent it! Do so as soon as possible. Vertical ventilation will delay heat build-up at the ceiling level of the burning room; it may delay flashover long enough to make possible a quick search for a victim, and it may also assist in the advancement of an attack hose line (Figure 4.8).

In some instances, however, *not* venting a burning room may also delay flashover and prove the best action to take. For example, if a burning room is filled with superheated smoke down to floor level and there are warning signs of rollover observed, don't vent the fire area; instead, close the door to the burning room. This action may delay flashover. By closing the door to a burning room in the pre-flashover stage, you may starve the fire of oxygen; thereby slowing down the rate of burning and heat generation necessary for flashover.

So, before you vent a window or door to a burning room, ask yourself the question: How is venting going to alter the fire condition in this room and affect the possibility of flashover?

A temporary slowdown to a flashover can be accomplished prior to stretching a hose by using a portable extinguisher on the fire. When discharged on the fire, the two and a half gallons of water in a portable extinguisher located on the premises or carried by firefighters can temporarily inhibit fire growth to flashover. This temporary postponement of flashover may be sufficient to save a life when there is a delay in hose stretching or a long, difficult hose stretch is needed. After the portable extinguisher is used and if the hose line is still not in position, however, the opening to the fire room should be closed, and the window should not be vented. Ventilation, coordinated with hose stream advance by an engine company, reduces the chances of injury due to flashover.

Non-Flashover Fires

There are some fires that do not flash over, in which fire growth is different. Fires in a very large area may not exhibit flashover. There may be no smoke build-up or sudden flash. If, however, the interior finish of the large area is combustible, flame spread may be extremely rapid, and firefighters cannot run faster than flame spread. Fire tests conducted on some plastic furnishings reveal a flame spread of as much as two feet per second. When a firefighter enters a large room to search for fire after penetrating more than five feet beyond an exit, the point of no return is reached. If a rapid flame front spreads along the under side of a combustible ceiling, the firefighter will not escape faster than the flames will spread.

Abandoned buildings whose windows and doors are all broken or missing may not flash over either. There is no confinement in which the smoke can build up. Again, however, rapid vertical and horizontal fire spread can trap a firefighter. The lack of compartmentalization provided by windows and doors in a vacant building speeds up fire spread even though it prevents flashover.

High-rise occupancies that have central air-conditioning systems or emergency smoke removal systems may not flash over because the air movement that results from the fans of the air systems will cause unusually rapid flame growth. This flame growth and fire spread will be unpredictable and will not travel to adjoining occupancies, according to scientific laws of convection, conduction, and radiation. Such unpredictable fire spread can trap a firefighter in a high-rise fire.

Lessons to Be Learned

1. The firefighter or fire officer conducting the initial entry and search for the location of fire is most endangered by flashover.
2. Most firefighters reported killed by flashover actually were disoriented and lost in smoke prior to flashover. The most effective defense against this flashover danger is an organized search procedure that helps a firefighter to search a smoke-filled room and return safely.
3. Fireground pre-planning that encourages firefighters to study floor plans and layouts can reduce the dangers of being trapped in a flashover. Condominiums, large housing developments, and town houses usually possess similar interior room arrangements. During training sessions, firefighters should analyze drawings of apartment floor plans. The knowledge of exit locations, windows, dead-end corridors, fire escapes, and stairways to second-floor bedrooms could give firefighters confidence during search operations and increase safety when they become disoriented in smoke.
4. When personnel are available, firefighters should search in pairs. If one firefighter becomes lost in smoke, the other firefighter may guide the disoriented firefighter to safety by voice directions; however, care must be taken when two firefighters

operate as a search team. There are dangers to the "buddy system" of searching; you must know the other firefighter on the team. There have been instances where one firefighter of a team was more interested in creating the impression of bravery than in safe search procedure. Searching with a firefighter who is primarily concerned with proving his daring can change a buddy system into a dangerous game.

5. Firefighters assigned to search a fire area should obtain search safety equipment that can be used to prevent disorientation in smoke:

    A *search light* should always be carried. In some instances, firefighters have fallen through floors and become trapped. Unable to return to safety, they were caught in flashover fires. A flashlight could have prevented this entrapment and death by flashover.

    A *search line*, 75 or 100 feet of small-diameter rope with a quick snap clip at one end, will also allow a firefighter to enter a large, smoke-filled area or a smoky high-rise floor area containing many small offices and to return safely.

    Self-contained breathing apparatus will prevent toxic smoke from overtaking a firefighter. Firefighters unconscious from carbon monoxide gas have been reported killed by flashover. Firefighters should not enter a smoke-filled room to search unless equipped with SCBA, with facepieces in place.

    Portable extinguishers are another protection against flashover. Even this small amount of water discharged on a pre-flashover fire may inhibit fire growth and provide a few minutes to save an unconscious victim or to escape an explosive flashover fire.

    The *voice* is another life-saving device. Firefighters who have been lost in smoke and have escaped just before a flashover tell of being led to an exit doorway by seeking the direction of another firefighter's voice.

    A *personnel alarm-sounding device* should be part of every firefighter's equipment. This electronic device, designed to sound a shrill alarm if a firefighter falls unconscious and remains immobile, can allow other firefighters to locate the unconscious firefighter quickly before a flashover occurs. In multi-story buildings, the danger of firefighters trapped on a floor above a fire is an ever-present danger.

    An *escape rope* that can be tied to a radiator pipe or other secure object and lowered out a window or over an outside balcony may allow a firefighter to climb down several stories below a spreading fire. This procedure is dangerous and should be undertaken only as a last resort, but an escape rope could prevent a firefighter who is trapped at a window by fire from turning to choose between burning or jumping.

6. Firefighters searching a room filled with smoke should plan to return via the same door through which they entered. The availability of the second exit in residence buildings can no longer be depended upon during a fire. Crime, or the fear of it, has locked the second exit in most homes (Figure 4.9).

Figure 4.9. *The fear of crime has locked up the windows and doors of many buildings and increased the danger of being trapped by flashover to firefighters performing search and rescue operations.*

Twenty-five or 30 years ago, firefighters, before entering a building, would size up the front of the structure to determine if the window leading to a fire escape, outside balcony, or a front-porch roof was available for a quick escape if the fire flashed over. If there was such an escape route at the front, a firefighter could consider taking the risk of passing fire to help a trapped victim.

This type of escape can no longer be depended upon. It has been secured and locked because of the possibility of a thief entering the apartment through such a window. First, sliding scissor gates were placed on windows, to prevent entry. Today, people are placing steel bars on the windows. These bars cannot be opened from the inside or the outside.

Another more deceiving barrier to using this type of window is the unbreakable plastic windowpane. Unbreakable plastic has replaced glass in windows that lead out to fire escapes, outside balconies, and porch roofs. A plastic windowpane cannot be broken quickly with a tool in order to escape from a smoke- or heat-filled room. A firefighter's axe or halligan tool will bounce off such plastic; the entire metal window frame must be removed. Firefighters will thus be trapped in a room about to flash over; quick escape by smashing open a windowpane when caught in a superheated room is no longer an option.

When searching in smoke, firefighters should always know the location of the door by which they entered. They should never let fire spread between them and that exit door. If that risk is taken and fire is passed to rescue a trapped victim, the firefighter should return through the entrance door with the victim.

7. Before firefighters in a company are assigned to search and rescue operations, they should be taught the characteristics of fire growth and learn to distinguish rollover and flashover. The first sight of rollover should be readily identified by all firefighters. The time available for a firefighter to escape after rollover and before flashover should be determined. Rollover precedes flashover; rollover must be considered a last warning to back out of a superheated fire area. Flashover signals the point beyond which no safety action or procedure can reasonably be expected to prevent the death of a firefighter in a burning room.

5

Advancing Initial Attack Hose Line Dangers

"ENGINE COMPANY 8, respond to the Armstrong housing complex, 126 South 5th Street. Building 'B'. The fire is reported on the third floor of garden-type apartment building."

"Engine 8, message received."

The captain records the information on a pad, then, nodding to the firefighter at the wheel, he switches on the electronic siren. The driver flips the switch to activate the emergency response lights, takes the apparatus on a sudden turn across a six-lane highway, and responds to the fire. The captain waves his arm out the cab window to gain the attention of motorists. Several miles later, the pumper turns off the highway and enters an apartment house complex. The officer takes mental notes: Six buildings, each about 40 by 100 feet. Three stories. Brick. Barrack-style "garden" apartments.

The captain leans forward in the cab to look for signs of fire. In the late afternoon sun, smoke is visible, drifting down from the roof soffit of one of the apartment buildings. The driver stops the pumper in front of the smoking structure and calls to the officer, who is leaving the cab, "Captain, we're in luck; there's a hydrant near the side of the building. I'm going to hook up to it." The officer nods, continues on to the back step of the apparatus, and orders, "Stretch from the 1¾-inch hose bed." The firefighters, adjusting their SCBAs and protective clothing, acknowledge the order.

The captain runs up a long path to the fire building, past a crowd of people gathered on the large grass lawn. They shout to the officer, "It's on the third floor! It's the 'B' wing—the right side of the building!" Glancing up at the windows, he sees that the top-floor windows on the right side of the building are discolored and blackened.

Suddenly, the portable radio strapped over his shoulder blares, "Engine 8 pump operator to Engine 8 officer. Captain, I have a dry hydrant. I'm going to hook up to a hydrant on the next street. The water will be delayed a few minutes."

"Engine 8, ten-four, but make sure we get water. We've got a top-floor job here."

The officer, climbing the entrance steps to the building, decides to take additional time to check the floor layout of the apartment below the fire and runs up the straight flight of stairs. There are two metal-clad fire doors at the top of the flight. The captain pushes one open. He moves 25 feet down a public hallway to two apartment doors at the very end.

Banging on one of the doors, he yells, "Fire Department!" An elderly woman opens the door to her apartment. "Lady, there is a fire upstairs; could I take a quick look at your apartment?" he says, gently pushing open the door. The woman moves aside, and he steps in to survey the room layout. There is a solid wall on the left. The kitchen and bedrooms are reached from a hallway approximately five feet farther into the apartment. The captain makes the quick assumption that the apartment next door must be a mirror image of this one. He turns to the woman. "Thanks. Say, ma'am, you'd better leave this apartment. There may be water or smoke drifting down here from the fire upstairs."

Out on the stairway again, he sees the lead firefighter coming up the stairs with the nozzle and hose draped over his shoulder. The sounds of the hose butts striking the metal risers and marble steps echo loudly. Firefighters stretching the line are shouting to the men below.

"We have a stairwell. Use it!"

"Chock that door open so the hose doesn't get caught under it."

The captain, who has ascended the second flight of stairs to the third-floor level, finds the metal-clad door in the corridor blistered and charred near the top. He pushes it open slightly, and dense black smoke and a blast of heat blow out; quickly he lets the door close. The captain dons the face mask of his SCBA and, once again, opens the door, but this time he crawls into the smoke-filled corridor toward the two end apartments. Visibility is zero; he cannot see the beam of his handlight. He can hear the crackle of burning dry wood; he feels the heat descending upon him. He is forced to retreat. Before turning back, he sweeps his hand widely over the floor around him on the chance of finding an unconscious victim. There are no bodies. He moves swiftly out of the superheated, smoky corridor on hands and knees and bursts out of the fire doors.

He pulls off his face mask. The nozzleman is bleeding the hose line of air pressure. The nozzle is cracked open slightly; air is hissing loudly out of the nozzle.

The captain says to the firefighters, "The fire apartment is straight down the hallway about 25 feet." The nozzleman nods his head.

Pusshhsss!

"A burst length!"

Someone in the stairwell shouts, "Captain, we have plenty of excess hose! We can remove the hose length and still make the fire apartment!"

"Okay, but hurry—this thing is really starting to cook."

The captain grips his portable radio. "Engine 8 to Engine 8 pump operator. Shut down the pumps. We have a burst length."

"Engine 8 pump operator, ten-four."

As one firefighter kneels on a doubled-over section of hose, another firefighter disconnects and removes the burst length and reconnects the hose line. The officer calls for the water over the radio.

Together again, the company is crouched down at the fire door. The line is charged. Everyone is once more wearing face masks, helmets, and gloves. The steady hissing of the positive-pressure masks drowns out the crackling of the fire. Down on one knee, the captain reaches in front of the nozzleman to push open the door and pats that man's shoulder. In one motion, the firefighter raises the nozzle tip upward to the flaming ceiling and pulls back the nozzle handle. A powerful stream of water sprays the upper portions of the raging fire in the corridor they are about to enter.

The firefighter crouched behind the officer quickly slips a wooden wedge beneath the door, chocking it in the open position. Entering the flaming hallway, bunched close together on their knees, the nozzleman, the captain, and the backup firefighter advance the attack hose line along the hall. Bursts of flame roll over their heads at ceiling level. The nozzleman quickly raises the hose stream upward, almost directly over their heads, to stop the flames. Back and forth, side to side, the hose stream whips frantically. The raging flames are beaten back.

Scalding water cascades down over the hose team. The flames begin to spread over their heads again, near the ceiling. The nozzle is used as a water curtain to block the fire's progress. Again, the flames are pushed back down the burning hallway. Slowly, crawling one foot at a time, the team moves forward. The nozzleman's head is down to avoid the radiated heat; he directs the hose stream ahead of him blindly. The captain behind him tries to look upward, searching the black smoke for signs of flame spreading over their heads. Suddenly, a heavy weight crashes on top of the officer's helmet, knocking him backward. His neck absorbs the shock—a plaster ceiling collapse. As he shakes his head, small red-hot pieces of plaster roll down his upturned collar. Despite this, the hose team stumbles several steps forward.

Crouched down on one knee, the hot concrete floor and hot ashes burn through the nozzleman's rubber boots. His knee, resting on the floor, is throbbing. The firefighter temporarily directs the host stream from the ceiling and down to the floor. He sweeps the hallway floor with water, but it doesn't cool down the area beneath him. He rises to both feet, crouched down in a duck walk. The hose line is again moved forward several more steps.

The hose team finally succeeds in reaching the end of the corridor. Two flaming apartment doors, side by side, are barely visible through the smoke. Both are partly open and flames show around their frames. The officer gently pushes the nozzleman toward the one at the right. The officer stretches out one leg in front of the nozzleman and kicks the apartment door further open.

Through the heat and dense smoke pouring out of the apartment, a faint glimpse of light from a window can be seen. The hose team crawls over the doorsill and enters the apartment. The officer shouts, his voice muffled through the hissing face mask, to the nozzleman right in front of him, "Go ahead about five feet and make a right turn!" They crawl forward, bend the hose to the right, and the heat in the apartment suddenly dissipates.

"Oh, God, we're in the wrong apartment!" the captain thinks.

Then he orders his team, "Quick! Back out! Back out! We're in the wrong apartment!"

They quickly pull the hose back out into the hallway. Flames are visible inside the adjoining apartment. They step several feet inside and sweep a 15-by-20 foot flaming living room with the hose stream. They move forward several more feet, turn to the left, and knock down the rest of the fire. The smoke and heat begin to subside; they can feel a cool breeze from a window. "I think we got it," the nozzleman says.

Several hours later, after the company has returned to the firehouse, the firefighters assemble in the kitchen to review the firefighting operation.

"Say, Captain," one of the firefighters says, "before you start we would like to say that we all agreed this was not one of our better jobs. Everything that could have gone wrong, did."

The officer raises his hand in protest. "Hold it—I disagree. I think it was one of our best firefighting efforts. True, we suffered three serious problems during the hose line attack—a dry hydrant, a burst length, and we almost wound up in the wrong apartment—but we overcame each one of these problems and went on to extinguish the fire. We overcame three setbacks and still triumphed.

"I've been at fires and seen fire companies come up against just one of these fireground problems and fall apart. That fire was never extinguished—the company could not overcome the setback. The attack hose line never extinguished the fire; a lot of yelling and shouting took place, and the blaze became a major alarm. We overcame three fireground problems, any one of which would have stopped an average firefighting company. I think it was a great job. It demonstrated our teamwork, training, and determination. It was in the best tradition of the fire service."

Advancing the initial attack hose line is a difficult tactic that exposes firefighters to many hazards. It is dangerous; it is also the most vital tactic in offensive interior attack strategy.

There are ten unexpected dangers that could kill or seriously injure firefighters advancing the initial attack hose line.

Rollover

The first danger faced by the attack team is rollover. If the door to an apartment on fire is partially opened upon arrival and a serious blaze involves two or more of the interior rooms, smoke and heat will be flowing out of the apartment door and into the public hallway over the heads of the firefighters (Figure 5.1). In the one or two minutes required for the firefighters in the hallway to bleed the uncharged hose line of air, adjust the face masks of their breathing apparatus, and play out the excess hose necessary for the advance into the burning rooms, the smoke and heat issuing from the doorway can suddenly burst into flame. It may necessitate retreat back to the stairway, and the firefighters will then have to battle the flames in the public hallway in order to regain the area from which they have just retreated. Their main objective is still to be accomplished: extinguishing the apartment fire.

This sudden flash of fire out of the apartment doorway is called "rollover," and it could have been prevented if one of the firefighters—

Figure 5.1. *Rollover is defined as sporadic flashes of flame mixed with smoke, coming out a doorway or window.*

after making a quick search to see if anyone were lying unconscious in the fire apartment—closed the door to the burning apartment.

Rollover is a flash phenomenon that occurs when combustible gases in the smoke and heat flow out of a compartmentalized burning area and mix with air, thereby entering the flammable range and suddenly igniting. The flaming gases, which are hotter than the surrounding air, cling to and flow across the ceiling in a "rolling" effect—possibly over the heads of the firefighters crouched down in the hallway waiting for the water to charge the hose line. Rollover occurs before the apartment flashes over.

Firefighters have been killed by rollover in multi-story buildings. If the hallway ignites into flame and a firefighter attempts to retreat but cannot find the stairs leading down to the street, and, instead, climbs up the stair leading to the upper floor, he will burn to death. When rollover occurs, the only safe exit is the down stairway.

Flashover

When firefighters crawl through several smoke-filled rooms to advance the attack hose line, the question on their minds is, "Will this thing suddenly light up on us—will it flash over?" The most reassuring sight a firefighter with a nozzle can see is the red glow of flame through the smoke (Figure 5.2). The most dangerous time of the hose

Figure 5.2. *After flashover has occurred, the danger of becoming trapped is reduced.*

line attack is the period when the firefighters are moving the hose stream forward through a superheated, smoke-filled hall or rooms without seeing flame. They must move toward the area holding the greatest amount of heat without knowing the size or location of the fire. It is during this time that flashover could occur.

The technical definition of flashover is full room involvement caused by thermal radiation feedback (re-radiation) from ceilings and upper walls which have been heated by the fire. Re-radiated heat gradually heats all the combustibles in a room to their ignition temperatures, and simultaneous ignition of the room contents occurs.

Each year, firefighters are killed by flashover; however, they are usually searching smoke- and heat-filled areas without the protection of a hose line. Hose line protection for firefighters, in which the water stream is directed for several seconds into the burning room, ahead of their advance, can inhibit flashover (Figure 5.3). The chance of a

Figure 5.3. *Water from an attack hose stream is the best protection against flashover.*

superheated, smoke-filled house or apartment flashing over and trapping initial attack team firefighters is not especially large. Rollover is a more common danger to an attack team than is flashover, which is one of the reasons why firefighters are advised to direct the hose stream at the ceiling in front of them as they move forward. Flashover has, however, trapped and burned firefighters advancing an uncharged hose line.

High heat is the indicator of flashover. If the heat build-up in a house or apartment is so severe that it requires firefighters to crouch down below a heat stratification level before entering from outside the fire area, the hose stream should be charged and even discharging water at the ceiling before they enter. Do not worry about water damage when there is a danger of flashover. Discharge the hose stream into the superheated smoke at ceiling level to inhibit any rollover or flashover before entering. Of the three firefighting priorities—life safety, fire containment, and property protection—the safety of the firefighter is included in the top priority.

Backdrafts

The so-called strip stores, taxpayers, or row stores suffer frequent backdraft explosions because they are difficult to ventilate. Fire in the center store of a taxpayer is in a tightly confined area. There are no side windows, and the rear door is usually difficult to force open quickly due to security precautions. If there is no skylight or if the structure is a two-story building, roof ventilation will be delayed or impossible; the only vent opening will then be the front entrance through which the firefighters are advancing the hose line.

At this type of fire, the situation can be compared to the firefighters crawling into the barrel of a loaded shotgun (Figure 5.4). If there is an explosion, they will receive the full force of the blast. The explosion could produce a shock wave forceful enough to blow them out of the store and on to the street; it could produce a ball of fire that severely burns them; it could cause a partition or ceiling collapse that might block the only exit and trap the firefighters inside the burning store.

A backdraft is an explosion caused by the combusion of a flammable gas-air mixture; it is triggered by the introduction of air into a confined space containing combustion gases, from a long-burning fire, that are heated to their ignition temperatures. Backdrafts generally occur in the main fire area several moments after the air has been introduced to the area—usually by the initial attack team that's advancing a hose line.

Backdraft Explosion in a Row of Stores

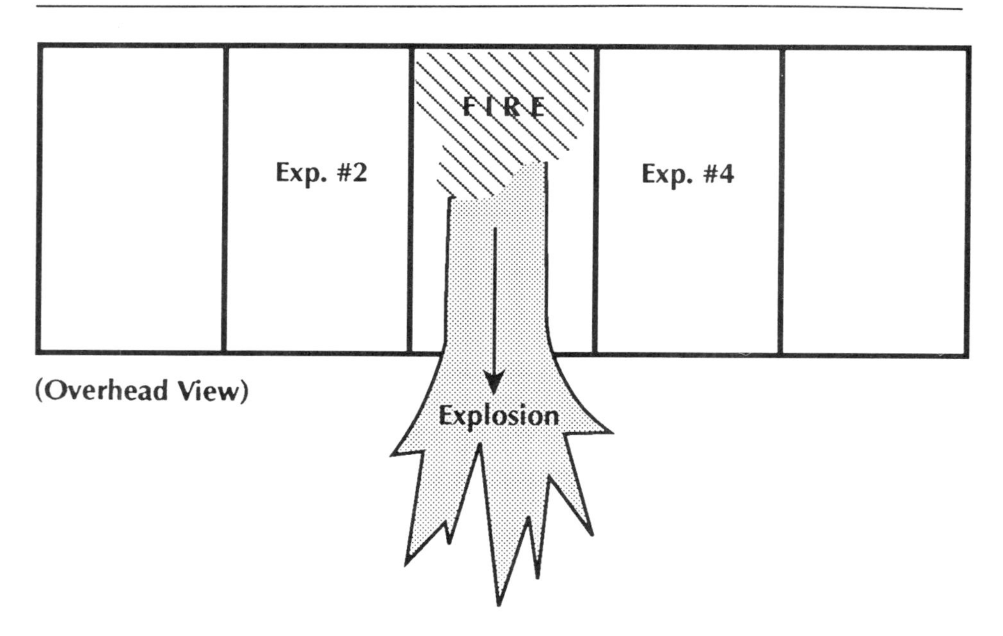

If there are no top or rear vents made in a burning store, firefighters, in effect, are standing in the barrel of a loaded shotgun about to fire when fresh air enters the store after forcible entry.

Figure 5.4. *At row store fires, advancing an attack hose line can be compared to crawling into the barrel of a loaded shotgun that is about to fire.*

When firefighters discuss the dangers of a backdraft explosion during training sessions, they invariably talk about warning signs. These warning signs include dense black smoke, smoke puffing around door frames, a reverse flow of smoke back into an open doorway, and discolored glass window. Experienced firefighters, however, observe all these signs at hundreds of different fires where backdrafts do not occur. Therefore, a firefighter should not believe that he can avoid a blast from a backdraft or from any explosion by observing warning signs (in fact, flammable vapor explosions, BLEVEs. and pressurized container explosions don't exhibit any warning signs).

Explosions happen too fast—there is no time to react. The only protection a firefighter possesses against an explosive blast is his protective equipment: gloves, mask facepiece, helmet earflaps or hood, turnout coat, pants, and boots. All exposed areas of the firefighter caught in the backdraft explosion may be severely burned: A firefighter wearing protective fire clothing and advancing a hose line should be able to take the full blast of a backdraft. If a firefighter is not in full fire gear with mask facepiece and gloves, and an explosion occurs, he will probably never extinguish another fire.

Overcrowding Behind the Attack Team

Many firefighters crowding behind a hose line attack team can lead to serious consequences:

- The additional firefighters could accidentally push an attack team member forward into a potentially dangerous position (Figure 5.5).

Figure 5.5. *Crowding up behind an attack hose line team inhibits an aggressive interior attack.*

- The proximity of these men could block or delay backward movement or temporary retreat to escape a blast from flashover or backdraft. Firefighters behind the attack team will actually be shielded from any blast of heat or flame; by the time they do feel it and reverse direction to retreat, the nozzle team will have been seriously burned.

- Crowding will inhibit an aggressive, forward-moving hose line attack. At a serious fire, a firefighter holding a nozzle will push aggressively forward down a burning hall when he knows there is room for a temporary retreat or knows he has the ability to stop the advance temporarily for several seconds, without being pushed ahead.

The area directly behind an attack nozzle team must be unblocked or guarded by only one or two company members; this practice will permit a temporary retreat if there is a blast of heat from a vented window on the other side of the fire or from an interior explosion. (The explosion dangers in a typical burning house are: gas accumulation from broken or melted stove pipes, vapors from an arsonist's flammable liquid, propane cooking or torch cylinders, coolant sealed in the tubing or refrigerant coils of a refrigerator, household pressurized containers, exploding window air conditioning units, kerosene containers used for space heaters, imploding television tubes, smoke explosions caused by double-pane insulated windows, and backdrafts.)

The advance of a hose stream at a serious fire is not a continuous forward charge. It is a slow, erratic forward movement, with short stops to cool down areas, and even short retreats along the way. The firefighter must have the psychological advantage of knowing his path to safety is clear.

Advancing a Hose Line Against the Wind

When the exterior wind velocity is in excess of 30 miles per hour, the chances of a conflagration are great; however, against such forceful winds, the chances of successful advance of an initial hose line attack on a structure fire are diminished. The firefighters won't be able to make forward hose line progress because the flame and heat, under the wind's additional force, will blow into the path of advancement (Figure 5.6).

Figure 5.6. *Firefighters cannot advance a hose line against a strong wind. Flame, heat, and smoke blowing into the path of the advancing firefighters will prevent forward movement.*

When the door to a fire area is opened and a strong wind is blowing from an opening at the opposite end of the structure, the actual point of fire origin may be several rooms back, out of reach of the hose stream. The flame blowing out the doorway will be the tip of the flame—actually the superheated gases igniting upon their mixture with air. A hose stream directed at this flame will have no effect, since the real generation of heat is several rooms removed from the flaming doorway.

To prevent serious burn injuries, firefighters encountering this situation must change their strategy. The interior line should be withdrawn and the door to the fire area closed. The officer in command must be notified of the inability to advance the interior attack hose line due to the strong wind. A second hose line should be advanced on the fire from the opposite end, the window or door through which the wind is blowing.

This method may require that firefighters stretch the line up an aerial ladder, fire escape, or portable ladder. The second attack hose line will advance on the fire from the upwind side. The initial attack team, behind the now closed door, must maintain their position to prevent the fire from extending into the public hall and to protect civilians and firefighters who may be using the interior stairs. Before the second attack team advances, the officer in command of that hose line must contact the first attack team to ensure that they have safely retreated behind the closed door and will not be injured by the heat currents and the steam created by the opposing hose line's advance.

Passing Fire

When a new firefighter in rookie school learns the fireground tactics of advancing an attack hose line, his company is told: "Never pass fire. It could cut off your escape and trap you inside a burning building." When the firefighter is assigned to a company, however, the fire chief and the officer sometimes order him to stretch the attack hose line to locations which require him to "pass fire."

At a house fire, for example, a company may be ordered to stretch the second hose line to the floor above the fire. To do so, they must pass the fire that's not yet extinguished by the first hose line on the fire floor. Perhaps the response is to a serious fire that has extended to the stairway of a three-story residence building. The captain orders the first line to "darken down" the fire, which started on the first floor, and then to proceed quickly up the flaming stairway to the top floor, extinguishing fire as they go. On the way up the stairs, however, a smoldering baby carriage in the hallway is passed; burning bags of garbage in the hall on another floor are passed.

Why were these objects passed? As an unqualified statement, "Never pass fire" is inaccurate; it should be rephrased: "Never pass fire which threatens to spread or increase in size to cut off your retreat or escape," or "Never pass fire if a second hose line has not been stretched when there exists a threat of fire spread or increase in size that could cut off your retreat or escape."

When ordered to stretch the first hose line at a stairway fire, the officer should request the chief to stretch a second hose line to back up the advance of the first. If, however, there is a delay with the second hose line and there is a serious threat of fire spreading out below and cutting off the attack hose line progressing up the stairway, the company officer should not expose the company to fire cut-off. He should, instead, wait for the backup line or leave a firefighter at the point where fire has been passed in order to warn the first attack hose line team of any potential fire spread and danger of being cut off. Likewise, when the second attack team is ordered to stretch a hose line to a floor above the fire, the officer of that company must visually check that the initial hose line is charged and controlling the original fire floor—before going up the stairs.

A fire officer should never allow his company to pass fire that threatens to cut off retreat. If the officer cannot accomplish the assigned task because of such a threat, the chief in command must be notified.

Burns from Collapsing Plaster Ceilings

Sections of plaster ceilings often collapse down on to firefighters as they advance the initial attack hose line. Large, thick portions of falling plaster ceilings can knock firefighters unconscious, cause a firefighter to drop the nozzle, and rain down hot pieces of plaster, melting paint, and sparks.

Ceiling collapse is caused by water absorption into the plaster from the hose stream. As firefighters move through a burning room or hallway, they direct the stream at the upper levels of the fire area to cool the atmosphere before crawling forward. While some of the plaster is broken and knocked down by the force of the hose stream, some of the water is absorbed by the remaining plaster ceiling. As the firefighter with the nozzle continues to advance, the water-soaked, red-hot plaster ceiling sometimes collapses suddenly, striking the firefighter on the head or crashing down on his shoulders, pieces rolling down the turnout coat collar or into the sleeves of the firefighter directing the nozzle upward. Some hot embers may even be caught by the top part of boots. These red-hot embers often burn the neck, wrists, and knees of the firefighter as he continues to move forward to extinguish the fire. If hot plaster from the ceiling is not falling on the firefighter, sparks, scalding water, or melted paint will probably do so.

To the firefighter advancing the hose line, however, these hazards come with the job—the real danger and concern is the unquenchable, roaring flames raging directly in front of him. If the firefighter understands and expects the likelihood of being burned by flaming substances from above, he can prepare for it by wearing proper protective clothing.

Floor Deck Collapse

When a mask-equipped firefighter crawls blindly forward in smoke, crouched down below the flames and heat, directing the nozzle stream at the ceiling overhead and being bombarded with hot pieces of falling plaster, paint drippings, scalding water, and sparks, he is depending on the floor to support his advance. He can endure the flame, heat, smoke, and falling hot objects and still move forward with the hose line, but he cannot advance if the floor shows signs of collapse. The firefighter requires a stable floor from which to mount a hose line attack and extinguish a fire (Figure 5.7).

The floor of a burning room can generally be depended upon to support the firefighters advancing the hose line; flames and heat rise,

Figure 5.7. *Firefighters must have a stable floor from which to mount a hose line attack.*

so the floor does not usually present a collapse problem. Over the past two decades, however, there has been an increase in fire-weakened floor deck collapse, and firefighters advancing hose lines plunge through with increasing frequency and sustain severe burns.

Increased incidence of floor deck collapse is mostly attributable to arson. An arsonist's flammable liquid spilled on the floor burns atop and under the flooring material. As the firefighter crawls forward, he may suddenly feel the floor deck give way, and one or both legs may plunge through it. If he is unable to pull his legs quickly up out of the floor holes or if other firefighters cannot free him, he will be seriously burned, straddling a charred floor beam or hanging between beams in the burning floor space.

To avoid this danger and advance a hose line safely through several burning rooms, crouching firefighters should keep one leg outstretched in front of them. Body weight will be supported by the back leg tucked beneath, while the outstretched leg can feel the stability of the floor deck or the presence of holes before further advance.

A serious fire raging below the flooring on which the firefighter advances a hose line poses a greater danger of floor beam collapse in addition to a floor deck collapse. The entire section of floor fails, dropping the firefighter down with it into the burning floor below. Bathrooms should be considered areas where floor collapse often occurs when fire on the floor below weakens their floor beams. Furthermore, firefighters should realize that a cement, ceramic tile, or terrazzo floor in a restaurant or kitchen may conceal a dangerously fire-weakened wood floor beneath it; they should know the size and intensity of the fire below these types of floors before advancing a hose line across them. Moreover, the areas around heavy machinery, stoves, and refrigerators will collapse first when the floor has been weakened by fire in a cellar. Firefighters advancing hose lines on a floor above a serious fire should avoid areas around heavy equipment. The reach of a hose line should be used to extinguish fire, while firefighters stay a safe distance away.

Incorrect Use of Master Streams

The powerful, high-pressure master stream directed through a window from a close-up approach can collapse an interior partition wall down on to firefighters; used inside a burning building, it can also explode several cinder blocks out of a wall that separates a

Figure 5.8. Firefighters operating interior attack hose lines have been injured by outside master streams.

burning apartment from a relatively uninvolved public hallway (Figure 5.8). A master stream directed into a burning room can create a scalding steam and hot water spray that can burn firefighters who are directing a hose line at a doorway. If directed at flame through an apartment window, the master stream can create a blast of superheated entrained heat and flame that can travel through several rooms. If it struck a firefighter in the chest, it could sweep him off a roof, knock him out a window, or hurl him down a flight of stairs.

A master stream is sometimes put into operation at a burning residential building when an interior attack hose team cannot advance on a fire. At some fires, the interior attack line cannot extinguish the fire because such areas as a fully involved attic or cockloft are inaccessible, or because of the large amount of fire present at such difficult structures as a factory, mill, or high-rise office, or because the wind is blowing into a window against the hose line's advance.

Over the years, firefighters operating interior attack hose lines have been injured by outside master streams. Sometimes the injuries are caused by the fire chief ordering the outside master stream into operation before receiving confirmation from the interior sector commander or company officer that all firefighters have been safely withdrawn; at other times, the injuries are caused by interior sector commanders or company officers who stubbornly refuse to change tactics and withdraw when the situation warrants it; sometimes the injuries are caused by the firefighter or fire officer who improperly directs the master stream without waiting for orders from the chief. Whatever the cause, poor firefighting tactics are almost always involved.

To set a master stream in operation safely requires coordination between three people: the chief in command outside, the interior sector commander inside, and the firefighter or officer in charge of the master stream. The chief in command must receive confirmation from the interior commander that all firefighters have been safely positioned or have been evacuated before starting a master stream in operation. The interior sector officer must have effective command and control over all of his firefighters and comply with the chief in command's orders to back out or withdraw interior firefighters to safety; the fire officer or firefighter in charge of the master stream must not direct it into the burning building until receiving orders to do so from the chief.

Incorrect Size-Up from Inside a Burning Building

Size-up of a fire can be accomplished from inside and from outside a burning building. An inside size-up is often made by the fire officer in charge of the initial interior hose line attack. The fireground commander, outside the burning building at the command post, will make the outside size-up. This commander will usually request first a size-up from the inside attack team, then make his outside size-up, and transmit a radio status report of the operation.

At the initial stage of a fire, the inside size-up is more accurate and useful than the size-up made from outside the building; the fire officers inside the structure are closer to the fire and, obviously, can see more of it than someone standing outside. Often when firefighters arrive at a fire located at the rear of a house or row of multiple

Figure 5.9. *When fire has spread to two or more floors of a building, the size-up outside the burning building at the command post is more accurate than the size-up inside the burning building.*

dwellings, no flame or smoke is visible at the front of the building. At such a fire, the officer making the inside size-up will discover the true seriousness of the blaze and report it to the chief. Conversely, at some fires the outside of a building may appear to be engulfed in flame and smoke, but the inside size-up will indicate that the fire is confined to one room and can be quickly extinguished with one attack hose line.

At most fires, inside size-up is more accurate. There is, however, one type of fire for which outside size-up is always more accurate, and failure to recognize the fact may cost firefighters their lives. This type of fire initially shows flame at only one lower or intermediate floor of a multi-story building, but, soon after the first attack hose line is advancing on the fire, the flames spread rapidly to several floors above. At the latter stages of this large-scale fire, the outside size-up will be more accurate than the inside size-up; in fact, size-up by the officer in charge of the advancing attack hose line may be dangerous incorrect. From his limited view inside, he may believe that the fire involves only one or two more rooms on his floor and can be quickly extinguished by the hose stream, when actually the entire building is involved in fire above the company and is possibly in danger of collapse (Figure 5.9).

This condition can occur when fire spreads up an open elevator shaft, through a vacant or partly demolished building with numerous holes in the floors, or in timber truss-roofed buildings. There have been several documented incidents in which fire chiefs have ordered advancing interior attack teams on lower floors to leave the hose and retreat from a burning building because fire spread rapidly to upper floors, threatening collapse, yet firefighters inside, unaware of the condition, did not obey the order quickly. Such incidents reinforce the importance of recognizing a fire that requires exclusively an outside size-up and of complying immediately with a chief's order to vacate a burning building.

Lessons to Be Learned

Fighting a fire by advancing an attack hose line down a hallway or through several flaming rooms of a house is dangerous (Figure 5.10), but it is also the most effective way to extinguish a fire. It saves the lives of trapped occupants and reduces property damage by preventing the spread of flame and smoke to adjoining buildings.

A hose line attack is the basic firefighting service provided by the fire department to the community. The youngest firefighters are usually the ones assigned to this critical firefighting task. Advancing a

10 Causes of Death and Injury to Firefighters Advancing an Attack Hose Line

1	**Rollover**
2	**Flashover**
3	**Backdraft Explosion**
4	**Overcrowding Behind the Attack Team Firefighters**
5	**Wind Blowing Fire in the Path of the Attack Team**
6	**Passing Fire**
7	**Ceiling Collapse**
8	**Floor Collapse**
9	**Master Streams**
10	**Incorrect Size-Up from the Interior of the Building**

Figure 5.10. *Ten causes of injury to firefighters advancing an attack hose line.*

hose line through flames requires their strength, determination, and courage. When the initial attack line can be rapidly advanced through the structure to the seat of the fire unimpeded, then containment, control, and extinguishment can be more quickly assured. If, through training and experience, this technique becomes standard operating procedure at structure fires, it allows everyone else to perform the duties with more safety and efficiency.

6

Operating Above Fire Dangers

"FIRE IS REPORTED on the ground floor of a 2½-story frame building at 142 West Street. People are trapped in the bedroom above the fire. Your response is three engines, two ladder companies, and a rescue company."

"Ladder 6 to Communications Center, message received. We are responding."

As the ladder company pulls up in front of the house fire, people dressed in nightclothes run up to the firefighters. A woman grasps the captain's hand and cries, "There's a fire in my house! My child is trapped upstairs in the bedroom! Please! Please! Save him!"

Heavy black smoke billows out of the partly opened front door. The windows have been broken by the bystanders. People are shouting to the firefighters running toward the building, "There are children inside!"

The chief arriving on the scene reports a working fire. He calls for another engine and ladder company to respond. The engine company lays a supply line from the hydrant to the front of the house. A preconnected 1¾-inch attack hose line is dragged into the front door. The ladder company captain and a firefighter don masks, enter the smoke-filled hall, and climb up the open stairway leading to the second-floor bedrooms. They reach the top of the stairs near the banister; the heat and smoke flowing up from below are severe. "Make a quick search and get out. This could light up any second," the officer announces in a muffled tone, speaking through the face mask.

He directs the firefighter to go to the bedroom on the left, while he searches the room on the right, directly over the fire. Smoke and heat have banked down to within two feet of the floor. The officer crouches low to get below the stratified heat and smoke level. His handlight is useless here.

Feeling the wall to his right, he enters the room and begins searching in a clockwise direction. He hears the crackling of burning wood from the room below. The beam of his handlight reveals smoke seeping up through the cracks between the floorboards. The officer stops momentarily to get around the piece of furniture he has just bumped into, head-on. He gets down on his hands and knees to avoid the heat.

The bed! The bed! Where the heck is it? He thinks as he moves around the room. He feels the edge of a carpet and the bed. He sweeps an arm under it, then runs both hands over the mattress. He calls out, "Hey! Anybody here?" No answer; no bodies.

The heat suddenly becomes worse. Smoke has banked down to the floor. Visibility is zero. The fire below is now a roar. There is still no sound of a hose stream striking the ceiling below. The captain decides to abort the search.

He crawls over the bed, back in the direction of the door. He crashes his head into a wall. No door! Feeling the wall in front of him for the doorway or a window, he moves to his left. He knocks over a small table and then bangs into a corner of the room. He changes direction and quickly moves to his right, blindly, in the smoke. He accidentally steps on the back step of a small tricycle, falls on his face, gets up with his boot tangled in the toy. When he attempts to free his foot from the bike, his boot comes off.

As the heat coming up from the floor below fills the room, he is forced to crawl on his stomach. His mask alarm starts to ring—air's running out. The captain comes upon a doorway and dives through it. His head strikes a wall, knocking off his helmet. He is in a closet. Clothes and boxes of toys fall on top of him. He grabs for his portable radio. "Urgent! Urgent! I'm trapped on the floor above. . . ."

The room flashed over!

One of the most dangerous assignments a firefighter can take is to go above a fire. Firefighters operating above a raging fire have been forced to jump for their lives from second-floor windows. They have been burned to death by blasts of flame as the room flashed over, their bodies discovered after the fire, hanging from open windows or crumpled up just below the sill.

The deadly products of combustion rise upward and kill firefighters trapped above a fire. Heat and flame may block their escape

Figure 6.1. *Flame and heat cause 25 percent of firefighter deaths.*

back down a stairway they have just climbed, or flames may quickly spread up the outside of the building from window to window. The smoke, heat, and toxic gases may seep through the cracks between floorboards, concealed spaces, and poke-through holes; a trapped firefighter may be asphyxiated by these products if his air supply runs out, or these products may react explosively in a fireball (Figure 6.1).

When a firefighter is killed above a fire, the cause of death may appear to be carbon monoxide or flashover; however, a careful analysis will reveal a deadly chain of events as the cause, not a single mishap.

First, the firefighter becomes disoriented. He is lost in smoke, entangled in some object, or confused by the sudden increase in heat or flame of the growing fire. Next, the firefighter is unable to return to the door or window he has just entered and is unable to find an alternate escape. Then he is overcome by smoke or toxic gases after the breathing apparatus runs out of air, he is burned to death by flashover, or he falls victim to hyperthermia, in which case his body adsorbs the rising heat faster than it can be evaporated.

Flame and heat cause 25 percent of the firefighter deaths in this country each year. Some of these victims are those trapped above a fire.

If operating above a fire is so dangerous, we must ask ourselves: Why do it? The answer is threefold: to search for trapped occupants of a burning building, to search for vertical fire spread, and to protect people trapped on a top floor during a shaft fire.

Searching Above for Victims

During a fire, firefighters must go above it to search for unconscious or trapped victims as soon as possible. In multi-story buildings, the most hazardous location, aside from the immediate fire area itself, is the room or apartment directly above a fire. Of the six exposed sides of a fire (the four sides and the top and bottom), the most deadly side is the one above.

A firefighter ordered to conduct a primary search of the fire structure will go above the fire as soon as possible after searching the point of origin. Deadly carbon monoxide is generated by incomplete combustion of the typical fire load in a burning house. The gas is lighter than air and quickly rises to the floor above.

Searching Above for Fire Extension

Immediately after a fire has been darkened down by an attack hose team, the chief in charge wants to know about conditions on the floor above: he wants to know if the fire has spread upstairs (Figure 6.2). A firefighter is ordered to go above to check for fire spread. This firefighter enters the room or apartment directly above the fire and feels the floor and baseboards there. If they are too hot to touch with the bare hand, or if smoke is pushing out around the cracks in the baseboard or flooring, then the area should be opened up and examined for hidden fire. Statistics reveal that when a second hose line is stretched into a burning building, it is most often operated above the fire.

Operating Above During a Shaft Fire

When a fire originating in a cellar or lower floor enters a shaftway that is enclosed or restricted at the top (such as a stairway, elevator, dumbwaiter shaft, or garbage chute), the flames quickly rise in the shaft, and the most rapid fire and smoke spread in the building will

Figure 6.2. Immediately after a fire has been darkened down by the attack hose team, the chief in charge wants a primary search and an examination for fire spread on the floor above the fire.

occur at the top floor. Here, the flames, heat, and deadly smoke, trapped by the top of the shaftway enclosure, will quickly mushroom out and spread to the top floor or attic.

People located at this level will be in great danger during the shaft fire. The chief in charge must order that the top of the shaft be vented to release the smoke and flame; at the same time, a hose line must be placed to extinguish the spreading fire, thereby protecting the top-floor occupants. To accomplish this task, firefighters must operate at least one or more floors above the fire.

If the fire breaks out of the shaftway enclosure on a lower floor, the firefighters could be trapped above. When, during a shaft fire, the chief orders firefighters to operate above it, the chief must also provide the means of escape, such as an aerial ladder.

Actually, some type of above-the-fire operation is carried out during every building fire. It becomes routine. When any procedure is performed over and over again, there is a danger of underestimating its seriousness. At every fire in which you are required to go above, a careful size-up should be carried out before you act.

Sizing Up the Fire

Most firefighters are trapped on a floor above a fire because they failed to size up the fire below them. The condition on the fire floor should be analyzed before going above. If not, a potentially deadly mistake is being made.

The firefighter should attempt to determine the approximate location of the fire. To check the hot spot above, the firefighter should know the hot spot below. Next, the size and intensity of the fire should be observed. (In most instances, only the flame and smoke coming from the doorway to the burning room or apartment can be observed.) This information should be used by the firefighter to determine if the fire can be extinguished by the hose attack team. If the fire appears beyond control of the firefighters operating the hose line, do not go above.

Sizing Up the Stairway Design

The type of stairway leading to the floor above must also be evaluated by the firefighter. There are three types of stairs: an open stair, an enclosed stair, and a smoke-proof tower stair (Figure 6.3).

An open stairway, found in most private houses, is the most dangerous stair a firefighter can climb. All the flame, heat, smoke, and toxic gases generated by the fire will flow up the open stairway leading to the second-floor bedrooms. An open stairway quickly becomes a chimney flue during a house fire. There is no protection for a firefighter attempting to gain access to the floor above a fire. In many instances, it is safer to go above a fire by way of a portable ladder placed at a second-floor bedroom window as a means of alternate or simultaneous entry.

The enclosed stairway of an apartment house offers more protection than open stairs in a private house. An enclosed stairway in a multiple dwelling that is equipped with properly operating self-closing doors to each public hallway may be used to go above a fire when the door to the fire floor is closed. When there are two such stairways, the safest way to go to the floor above is by way of the one that is not being used by the attack hose team to extinguish the fire.

Types of Stairways

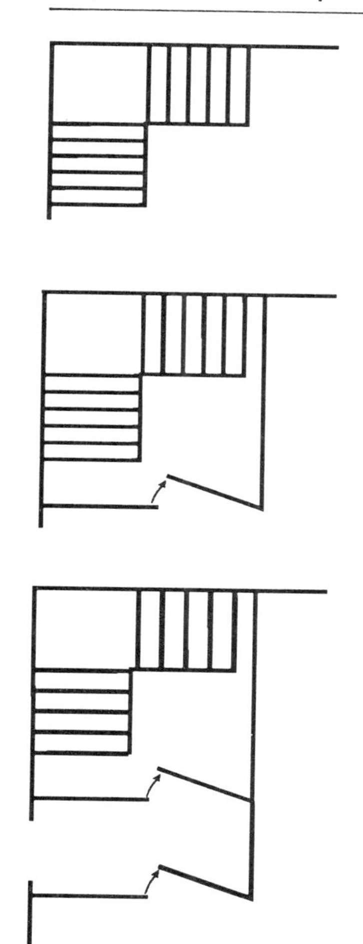

Open Stairway

An open stairway found in a private house is the most dangerous stair to climb when searching above a fire. It becomes a chimney flue.

Enclosed Stairway

An enclosed stairway offers more protection only when the door to the fire apartment is closed.

Smoke-Proof Tower Stair

A smoke-proof tower stairway is the safest stairway for a firefighter to use when searching above a fire. An open air vent exists between the floor occupancy and stair enclosure. Smoke will not enter the stair and trap a firefighter on the floor above.

Figure 6.3. *There are three types of stairways in a building: an open stair, an enclosed stair, and a smoke-proof tower.*

When an apartment house possesses only one enclosed stairway and the door to the burning apartment is open during the fire attack, the enclosed stairway will become filled with heat and smoke. Before going above the fire, the firefighter should check to see that the hose line is being advanced into the burning apartment by the hose attack team (Figure 6.4) and that the door to the burning apartment has not been removed or damaged beyond use by the forcible-entry operations. If the attack team is compelled to retreat by a burst hose, an explosion, or an increase in fire intensity, the door must be closed to protect the firefighter above.

***Figure 6.4.** Before going above a fire, the firefighter should ensure that a charged hose line is being advanced by an attack hose team and is extinguishing the fire.*

Some commercial high-rise buildings are constructed with a smoke-proof tower stairway, in which an open airway between the occupancy and the stair enclosure prevents heat, smoke, and flame from entering the stairway. This type is the safest stairway for a search above the fire.

Sizing Up a Second Exit for Escape

If the interior stair used by the firefighter to go above a fire suddenly becomes filled with heat and flame, the firefighter cannot use this path to get back down. He must locate a second exit for his emergency escape or be trapped above. He must know his options ahead of time, should the worst occur. Going above a fire should not be a snap decision made on the spur of the moment inside the burning building. It should have been decided upon at a pre-planning session, at the start of the tour of duty back in the firehouse.

With this knowledge, the firefighter can on arrival properly size up the outside of the burning building. Before entering a burning

Figure 6.5. Firefighters should identify a second exit before going above a fire. A portable ladder should be raised to a window on the floor above the fire.

building to search above the fire, he should examine the front of the structure and look for a second exit. A portable ladder already raised to a second-floor bedroom window (Figure 6.5), a porch roof, or a fire escape may provide an escape if the interior stairs become impassable because of fire.

Building Construction Size-Up

The degree of danger or threat of being trapped above a fire is greatly influenced by the construction of the burning building. Of the five basic types of building construction—fire-resistive; non-combustible; ordinary (brick-and-joist); heavy timber; and wood-frame—the greatest threat to a firefighter who must search above a fire is posed by the wood-frame building (Figure 6.6). Vertical fire spread is more rapid in this type of structure. In addition to the three common avenues of vertical fire spread—the interior stairway, windows (auto-exposure), and concealed spaces—flames can trap a firefighter above a fire in a wood-frame building by spreading up its combustible exterior. No other construction type offers a combustible exterior. In

Construction Type and Degree of Danger When Searching Above a Fire

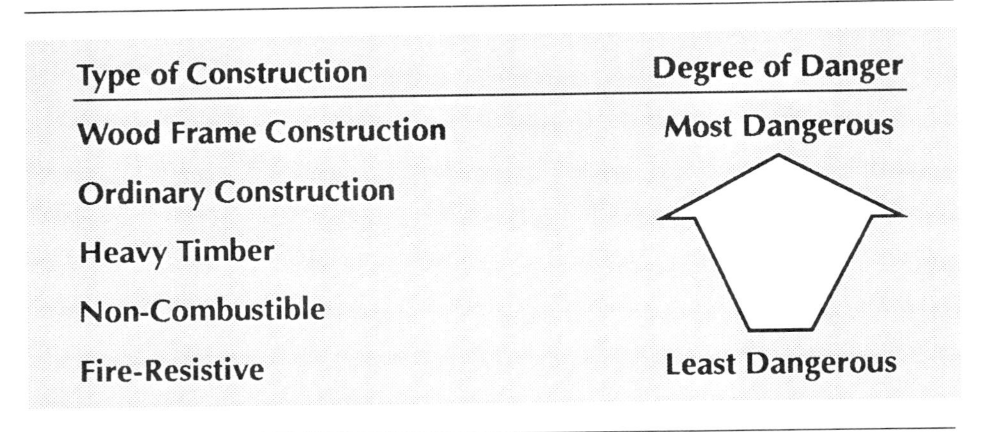

Figure 6.6. *Searching above a fire in a wood-frame building is extremely dangerous. Rapid fire spread on the inside and the exterior can cause a firefighter to become trapped.*

addition, interior walls, halls, and stairs of wood-frame buildings are often covered with combustible wood that rapidly spreads fire upward to the floor above. Firefighters who transfer from a fire district with brick-and-joist buildings to a fire company that responds to fires in wood-frame buildings must realize that they have less time to search above a fire before the danger of being trapped becomes great.

A building of ordinary (brick-and-joist) construction is the next most dangerous structural type in which to operate above a fire. The vertical fire spread problem in this type of structure is caused by the concealed spaces. These spaces, behind walls, ceilings, and floors, spread flame and smoke upward to the floor above. Small openings, cracks around radiator pipe risers, and ceiling light fixtures allow flame and smoke to enter and emerge from these concealed spaces.

Many communities and towns include areas in which buildings of different construction types are side by side. Firefighters must realize that the dangers in operating above a fire depend, not only on the size of the fire, but on the type of building construction as well.

Sizing Up the Fire Floor Operations

A firefighter going above a fire must be able to size up accurately the hose attack team operation taking place on the fire floor. Before going above, he should first check to see that the hose line is charged and all is in place for commencing the initial attack operation.

The firefighter about to advance the attack hose line must be wearing full protective clothing and self-contained breathing equipment (Figure 6.7). The hose attack team should be just about to enter the fire doorway or be already inside the fire area, while the charged hose line is being advanced with them. The scene should not indicate a disorganized defensive operation in which water is shooting into a smoke-filled doorway, but, rather, an aggressive, well-planned interior attack on the origin of the fire with fully protected firefighters.

If the attack team is about to advance, the officer in command of the hose line should be informed by the firefighters responsible for the above-fire operations that he is going above. This communication should make the attack hose line officer realize that more is at stake than simple fire extinguishment. Moreover, the communication also makes him responsible for notifying the chief in command of the upward-bound firefighter's precarious position should the hose line advance be unsuccessful.

Firefighters have been trapped above a fire even after conducting a proper size-up of the fire floor operation. Unexpected events often happen at fires: A pumper may suddenly break down and cease

Figure 6.7. *Make sure that firefighters who are advancing the attack hose line are wearing full protective clothing and self-contained breathing apparatus and are aggressively moving into the fire area before other firefighters go above the fire.*

pumping water to the attack hose line; a hose length may burst and delay the hose line attack advance; a window in the burning room may break, allowing a strong wind to blow fire against the advancing hose team, forcing them back out of the burning apartment; an explosion may seriously injure the firefighters operating the attack line. Therefore, firefighters must realize that working above a fire is always a high-risk operation.

Lessons to Be Learned

After the decision has been made to go above a fire, there are several safety precautions that a firefighter can take to reduce the risk of being trapped.

1. Notify your officer when you go above a fire. Even if your assignment has been pre-planned, inform him by portable radio. This information is a form of fireground control that increases firefighter safety. A company officer should know where all of his assigned firefighters are operating during a fire.
2. When the officer in charge of the attack hose line is crouched down in the hallway, about to advance the hose line into the burning room, a tap on his shoulder and a finger pointed upstairs can convey your assignment to him.
3. When firefighters are searching in pairs, leave a firefighter at the foot of the stairs to warn you of deteriorating conditions on the fire floor.
4. If numbers permit, one or two firefighters should be assigned to assist an undermanned or inexperienced attack hose team on the fire floor.
5. When there is a danger of flashover in the hallway above a serious fire and a difficult forcible-entry operation is required to gain access to the apartment directly over the fire, locate an open door or force open a door to an apartment that is not directly above the fire. If conditions suddenly grow worse in the hallway, the open apartment which is not over the fire may be your area of refuge.
6. When the apartment or office floor above the fire in a fire-resistive building is opened and the area proves to be clear of smoke and heat, but the hall is charged and about to flash over, it may be feasible to close the door to the apartment partway after you enter to search, so that flame or smoke does

not spread inside and trap you. In many older residence buildings, however, the smoke and heat will be worse inside the apartment directly over the fire apartment. They seep upward through concealed spaces and poke-through holes. Do not close the door behind you when searching this type of smoke- and heat-charged area above a fire.

7. Make sure you are equipped with a portable radio and full protective fire gear. If you become trapped above, notify the officer in command of your condition and your location.
8. If you enter a smoke- and heat-filled room, hallway, or apartment above a fire and you suspect imminent flashover conditions, first locate a second exit, such as a window leading to a fire escape or a portable ladder, and then start to search (Figure 6.8).
9. When you climb or descend a stairway between the fire floor and the floor above, stay close to and face the wall. Heat, smoke, and flame flowing up a stairway will rise vertically near the stairwell or around the banister.

Figure 6.8. *When you are above a serious fire, locate a second exit first and then begin your search.*

10. All firefighters assigned to search above a fire should understand the firefighting priorities of risk-taking (Figure 6.9). The only justification for risking a firefighter's life is present when there is a real chance to save another person's life. If someone's cries for help can be heard, or if a victim is observed lying on the floor, a firefighter may take any chance in an effort to save that person, including sacrificing his own life. A firefighter should not, however, risk his life on the report of a missing person or even the high probability of a person trapped above a fire. We have all witnessed the person shouting about "my baby" being trapped—and the baby turns out to be a cat or the hallucination of a hysterical person. When such a vague plea for help is heard, the veteran firefighter stays calm and is not carried away by the emotional scene but makes every effort humanly possible to search the area, short of becoming trapped and killed by the fire.

Figure 6.9. *Firefighters are forced to crash through the nearest window when rooms fill with superheated gases or flame.*

Appendix to Chapter 6

An analysis of firefighter deaths in the FDNY over the past 30 years reveals 16 officers and firefighters have been killed while operating above a fire.

FDNY firefighters who have died operating above a fire:

Lieutenant Richard Mac Clave	November 26, 1957
Captain Erick Thomas	April 3, 1959
Firefighter John Crosthwaite	February 4, 1961
Firefighter Robert Hurst	November 22, 1961
Firefighter Charles Lang	November 22, 1961
Firefighter John King	December 27, 1961
Lieutenant Eugene Miller	December 24, 1968
Captain John Dunne	March 28, 1971
Firefighter Henry Mitchell	August 31, 1972
Lieutenant Joseph Sparacino	August 30, 1976
Firefighter Martin Celic	July 10, 1977
Firefighter Larry Fitzpatrick	June 27, 1980
Firefighter Gerard Frisby	June 27, 1980
Firefighter Philip D'Adamo	December 1, 1984
Firefighter Robert H. Dayton	November 26, 1988
Firefighter John P. Devaney	February 3, 1989

7

The Peaked-Roof Dangers

TWO FIREFIGHTERS carry a roof ladder toward the burning house. Flames are shooting out of the front attic window. Approaching the house, the lead firefighter stops momentarily, lowers the ladder from his shoulder, pushes the spring-loaded hooks on the end of the ladder inward, then turns them at a 45-degree angle until they lock open.

Replacing the ladder on his shoulder, he and his fellow firefighter start to climb up another ladder to the roof of the burning building. They slide the roof ladder along the sloping surface of the peaked roof; the hooks are secured over the roof's ridge. Climbing up the roof ladder, they step on the rungs to keep from sliding down the slope. One firefighter carries an axe, the other a pike pole.

"Start water!" a firefighter in the front yard below them yells to the pump operator. The hose stretched into the doorway suddenly jumps off the ground as the water pressure surges through it.

Crash! Crash! Heavy black smoke suddenly flows out of the top-floor windows that a firefighter has just broken.

The firefighter with the pickhead axe makes his way across the roof, straddling the roof ridge as he walks, while his partner with the pike pole waits at the ladder. The firefighter above stops at a brown heat stain on the roof and starts to pry up the asphalt shingles and roofing paper with his axe. Smoke pours upward between the roof boards as they are exposed.

He carefully takes several steps down the slope of the roof and raises the axe above his head to begin cutting the roof. Suddenly, his right leg plunges through the fire-weakened roof deck. As the firefighter loses his balance, the axe flies out of his hand. He falls backward. His back and head slam into the roof, knocking his helmet off; it rolls down the roof. He's flat on his back with his head toward the edge of the sloping roof—only his right leg, stuck in the smoking roof hole, keeps him from sliding off the roof. He tries to sit up, but the weight of the mask cylinder on his back makes that impossible.

"Help! Joe! Get me out of here!" he yells to his partner. He feels the heat of the attic fire on his right leg; if he pulls it out of the hole, he may roll off the roof. Again he tries to sit up. The heat from the burning attic directly below him is being conducted upward through the roof deck, burning his back. Because his leg is hooked into the roof, he feels pain in his twisted knee, and his foot is growing hotter. He tries again to lift himself up out of the hot asphalt shingles; his gloves push down into the hot melted tar. It's no use. His head slams back down on the roof.

Almost upside down, looking up into the black night sky, he feels something hard hit his chest. He looks up through the smoke and sees his partner standing on the peak of the roof, holding an outstretched pike pole down toward him. Quickly he grabs the metal ends with both hands and is pulled upright. In a crouch, holding on to the hook, he pushes down on the free leg to pull himself up out of the collapsed roof deck hole.

Suddenly, his other leg plunges through the roof. He loses his grip on the pike pole and falls through the burning roof up to his waist.

"Help! Joe!" he cries again. Smoke and heat rise up around him. Holding on to the roof with outstretched arms to keep from falling completely through the roof hole, he feels the deck around him crumbling. His legs move beneath him as he desperately seeks to find an object to step on so that he can boost himself up from the hole. There is nothing. He feels pain in his lower torso from the heat of the fire below.

Suddenly he feels himself being pulled upward. His partner, standing above him, has grabbed the backplate frame of his SCBA and is pulling for all he's worth. Now halfway up out of the roof, he is able to grab the roof's ridge. With a last-ditch effort and the help of his partner, he pulls himself up out of the burning hole.

Half crawling, half being dragged, he at last arrives back at the roof ladder. His partner calls, on the Handie-Talkie,™ "Roof team to command post. We need medical assistance. There is a burned fire-fighter. There was a roof collapse!"

The most dangerous roof on which to operate is a peaked roof. The hazards of such operations have been greatly underestimated. A sloping peaked roof requires a firefighter to perform acrobatic feats of climbing and balancing while simultaneously carrying out urgent firefighting tactics. Most fires in America occur in residential buildings, and most residential buildings have peaked roofs.

There are five types of peaked roofs found on houses throughout this country. A *gable* roof is the most common: it has two sides sloping up from two bearing walls and meeting at the top ridge rafter (Figure 7.1). A *shed* roof slopes upward from only one side and is supported by two bearing walls of different heights. A *hip* roof has four sides sloping up from four bearing walls (Figure 7.2). A *gambrel* roof, often found on barns, has two slopes on each of two sides, with the lower slopes steeper than the upper (Figure 7.3). A *mansard* roof, named after French architect François Mansart, has two slopes on each of four sides; the lower slopes are steeper than the upper (Figure 7.4).

A peaked roof (Figure 7.5) is more dangerous than a flat roof in several ways:

- There is no stairway, fixed ladder, or adjoining building that can provide safe access to a peaked roof—fire department ladders are always needed.
- There is no parapet around the edge of a peaked roof to keep a firefighter from walking off or falling off at night.
- The surface of a sloping peaked roof, covered with slate, tile, or wood shingles, is too slippery to walk on when wet, icy, or coated with wet leaves or moss.
- A peaked roof is built to support less weight than a flat roof because it has been designed to shed snow. Peaked roofs sometimes have flimsy two-inch by four-inch roof beams, and the roof deck sometimes consists of furring strips nailed to the joists, two or four inches on center. This type of roof deck serves only to fasten shingles—it will not support the weight of a firefighter.

Despite these dangers, firefighters must operate on peaked roofs to overhaul roof shingle fires, extinguish chimney fires, wet down wood-shingle roofs for protection against nearby exposure fires, and cut roof vent openings. You can follow some safe operating procedures to help prevent you from losing your balance, sliding or rolling off a roof, or plunging through a burned-out roof deck, and becoming trapped in a burning attic.

Figure 7.1. *A gable roof has two sides sloping up from two bearing walls, meeting at the top ridge rafter.*

Figure 7.2. *A hip roof has four sides sloping up from four bearing walls.*

Figure 7.3. *A gambrel roof has two slopes on each of two sides, with the lower slope steeper than the upper.*

Figure 7.4. A mansard roof has two slopes on each of four sides; the lower slopes are steeper than the upper.

Typical Peaked-Roof Construction

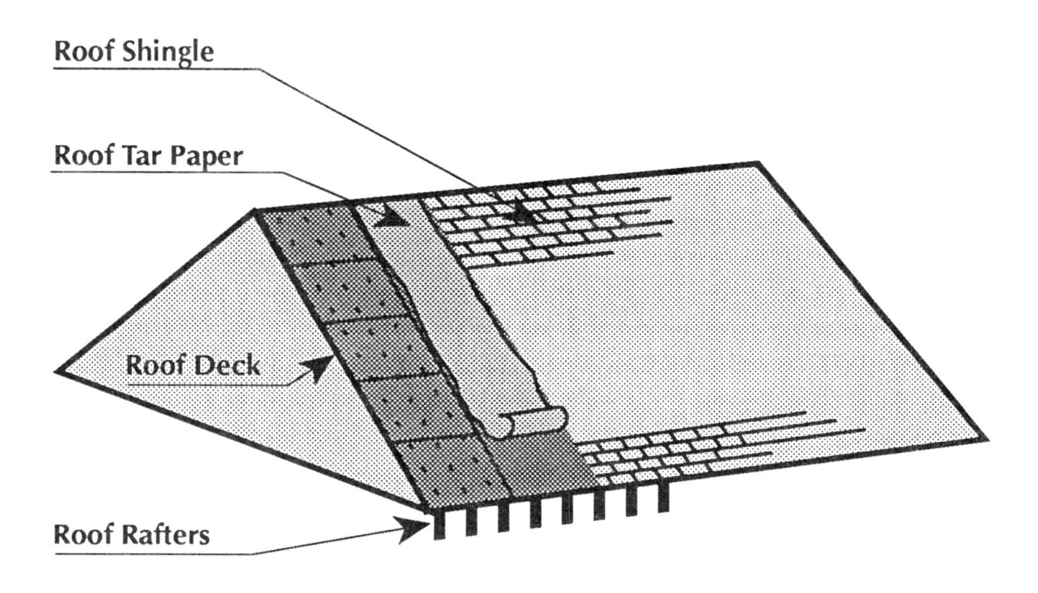

Figure 7.5. A peaked roof consists of roof rafters, roof deck, roofing paper, and roof shingles.

Sizing Up the Dangers

Before a firefighter climbs on to a peaked roof to operate during a fire or emergency, he must conduct a roof size-up to evaluate the four common dangers associated with a peaked roof: the roof deck (the layer of wood surface between the supporting roof rafters and the roof shingles), the roof rafter support system, the roof slope, and the roof shingle surface.

Roof Deck

A burned-out roof deck is the most common type of peaked roof collapse danger. A firefighter walking on what appears to be a sturdy roof with a new shingle covering can suddenly plunge one leg through a fire-weakened roof deck. Then, if he uses the other leg to push himself up out of the hole, he can cause the collapse of a nearby section and sink deeper into the crumbling roof. He will either be caught up to his waist or will fall through the roof deck between the rafters and be trapped in the burning attic.

To protect yourself from such a roof deck collapse (Figures 7.6, 7.7, 7.8, and 7.9), use a portable roof ladder. The safest possible

Figure 7.6. *Firefighters supported on a peaked roof by a roof ladder.*

Figure 7.7. ABOVE: *A firefighter steps off the roof ladder and falls through the fire-weakened roof deck.*

Figure 7.8. BELOW LEFT: *Another firefighter, attempting to help, also falls through the collapsing roof deck.*

Figure 7.9. RIGHT: *One firefighter pulls himself up from the collapse hole.*

precaution is to have a roof ladder long enough so that, when the hooks are secure over the ridge, the base of the ladder reaches to and rests on the bearing wall. Then, if the roof deck collapses, a firefighter standing on the rungs of the roof ladder will be safely supported.

Modern building codes require solid roof decks of ½- to ¾-inch plywood or tongue-and-groove boards. This type of deck will provide some stability for those operating on a peaked roof during or after a fire. If, however, the stability of the roof deck is in doubt, operate from the rungs of a portable roof ladder.

Roof Rafter

At some fires, the entire roof collapses, not just a portion of the roof deck. The roof rafter, the ridge rafter, the portable roof ladder, and all the firefighters on the roof collapse into the top-floor or attic fire. A portable roof ladder, hooked on to the roof ridge and supported by the bearing wall beneath the eaves, will not protect a firefighter when the entire roof rafter system collapses during a fire.

To operate safely on a peaked roof, a firefighter must know the type of roof rafter support system that is holding up the roof. The three most common types of roof rafter systems used to support peaked roofs are timber truss, plank-and-beam, and rafter construction spaced 16 inches on center (Figure 7.10).

From a collapse standpoint, the most dangerous roof rafter system is the truss. If one truss suddenly collapses, a large area or the entire roof may collapse as well. One single timber truss can support one-third of the roof. Firefighters are accustomed to the collapse of only a small portion of a roof during a fire—not such a large area of roof deck.

Timber trusses may be spaced as much as 15 or 20 feet apart. Most of the roof will therefore be unsupported. Whether a firefighter operating on a peaked roof plunges through a burned-out roof deck and becomes trapped in the fire below depends on the spacing of the roof rafters that support the roof deck. Most of the roof above a truss is unsupported plywood deck or tongue-and-groove one-inch board. If the roof collapses, there is little chance for a firefighter to grab a truss or purlin (a longitudinal beam connecting the timber trusses together).

A plank-and-beam system provides the next largest area of unsupported roof deck. The safest roof support system on which to

Roof Deck Collapse Danger Depends on the Type of Roof Support System Used Beneath the Deck

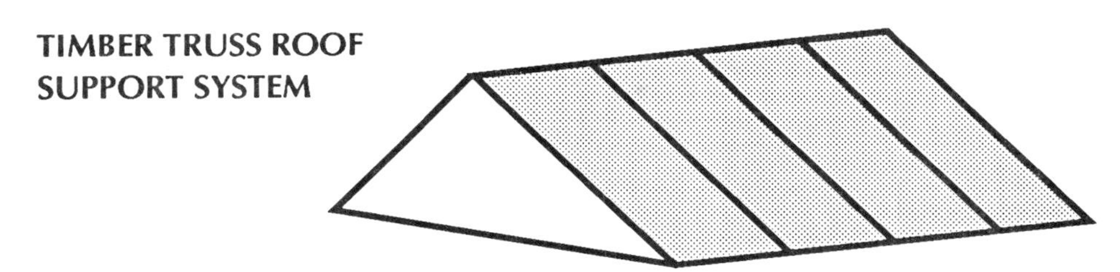

On a truss roof most of the roof deck above the truss can be unsupported plywood. There is little chance to grab on to a truss if the roof deck burns through.

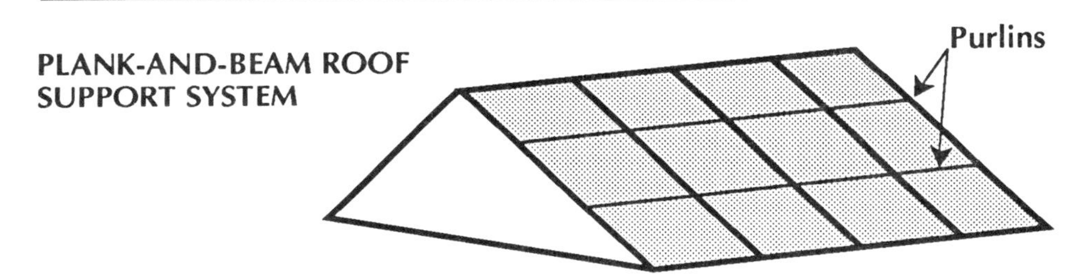

A plank and beam roof support system is similar to a truss except it has purlins. These are longitudinal beams connecting the timber trusses together.

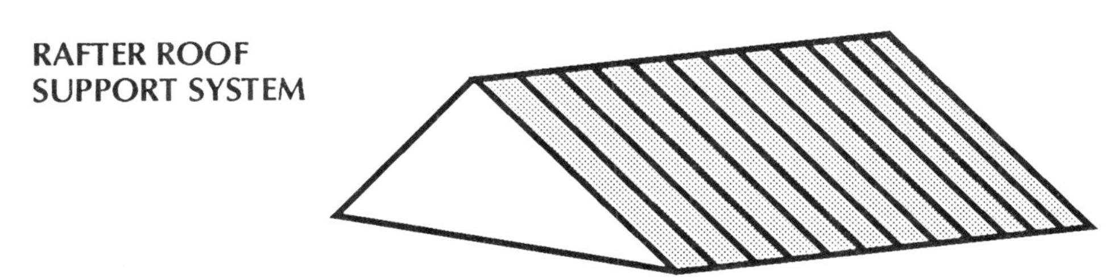

The rafter roof support system's close spacing gives a firefighter a better chance of grabbing on to roof support if the roof deck collapses during a fire however; a firefighter can fall between rafters into a fire.

Figure 7.10. *There are three types of peaked roof construction: the truss peaked roof, the plank-and-beam peaked roof, and the rafter peaked roof.*

operate is a rafter system spaced 16 inches on center; if such a roof deck gives way beneath a firefighter, his chances of becoming hung up or grabbing a rafter are good. He may then be able to pull himself back up out of the burning hole in the deck.

A firefighter might ask, "How can I possibly know the type of roof rafter system or roof deck beneath a shingled peak roof?" The answer is that you had better know this information. You can find out by studying your local building code, by inspecting buildings during the construction phase, and by examining the roof supports and decks from the under side every time you overhaul an attic or top floor after a fire. If you don't know what the roof construction of a building is, you'd better not stand on it during a fire.

If you are responding to a fire, are ordered to operate on a peaked roof above a serious attic blaze, and do not know the type of construction supporting the roof, raise an aerial ladder or aerial platform to the roof and work from the safety of an independently supported ladder (Figure 7.11). Even if the entire roof suddenly collapses, you will still have support. If, however, you cannot position the aerial on the roof due to obstructions, if you do not know the type of roof construction, and if you nevertheless climb up on the peaked roof, you must realize that you risk plunging through a fire-weakened structure into a dark, smoky attic or top floor.

Figure 7.11. *Operating atop an aerial ladder is the safest way to vent a peaked roof above an attic fire.*

Roof Slope

The chances of a firefighter losing his balance and falling off a sloping rooftop are greater than those of a firefighter being injured in a collapse. In roof operations, firefighters are sometimes required to leave the safety of the ladder and walk on the sloping roof surface. On a low-pitched roof (one which is at a 30-degree angle or less from the horizontal), a roof ladder may not be required, as the risk of losing one's balance due to the slope is not great.

There is, however, a safe method of walking on a sloping roof that all firefighters should know: the flat-footed method. Do not walk using a normal heel-toe foot action. Moreover, in order to shift your body weight quickly and to compensate for the uneven surface of the roof, bend both legs at the knees. You should not walk straight down the slope of a roof but rather walk across it, at an angle.

Guidelines for Ladder Use on Peaked Roofs

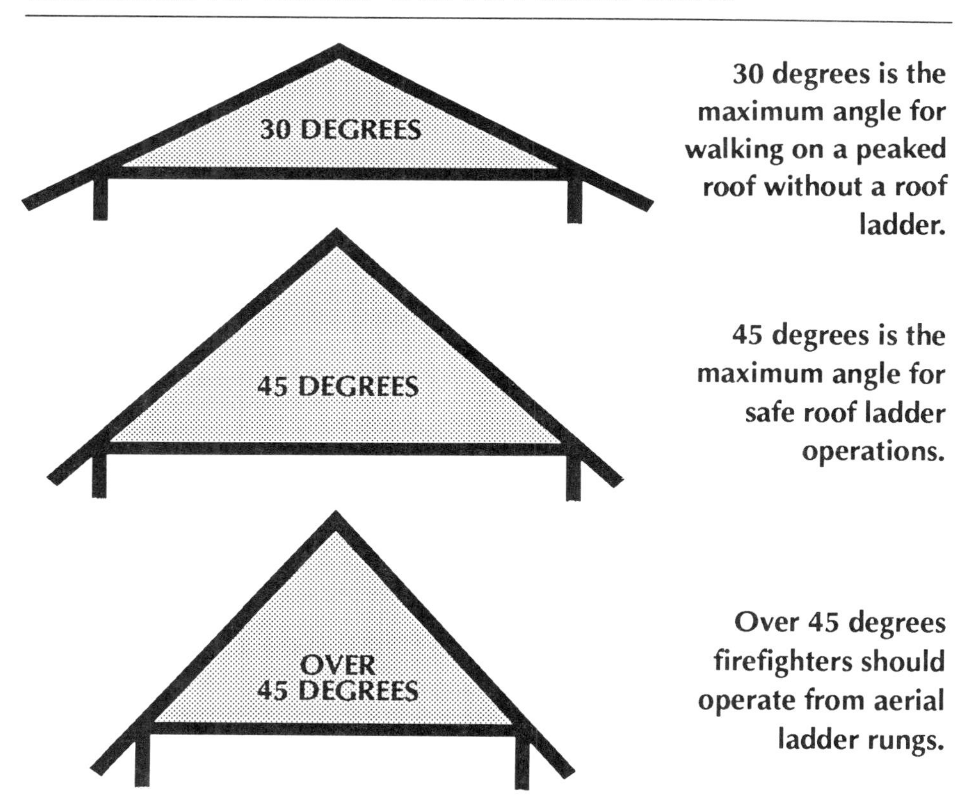

Figure 7.12. *Roof slopes that are more than 30 degrees from the horizontal are too steep to walk upon.*

Walk toward the edge of a roof very carefully. At normal walking pace, the weight of a mask and protective clothing (almost 50 pounds) can increase your forward momentum and prevent you from stopping at the edge of the roof.

Roof slopes of greater than 30 degrees are too steep to walk on. You need a portable roof ladder to move up and down such a sharp slope (Figure 7.12). If it is absolutely necessary to leave the ladder, walk straddling the ridge. If you trip or lose your balance, grab the ridge of the roof—it is your only sure handhold. Do not rely on chimneys, television antennas, and soil pipes; they may collapse under your weight.

Roof Shingles

Always size up the surface of the peaked roof before climbing on to it. Even if the slope of the roof is small, the surface may be too slippery to walk on.

Slate and tile shingles are more dangerous than asphalt shingles. Slate is extremely slippery when wet; tile shingles can crack when you walk on them, slide out from under you, and cause you to fall. Even asphalt tile, which offers fairly good traction, can become heated by fire and melt, turning into a slick, oily surface.

Wood shingles sometimes develop a fungus or moss near the bottom edge of the roof deck when the house is located near a lake or ocean. Furthermore, avoid any type of shingle surface located directly underneath a large, overhanging tree. Sap, leaves, or seedlings dropping from the tree will create a dangerous roof surface. Any type of set sheet-metal roof will cause a firefighter to lose his balance no matter how gentle its slope.

Some roofs in northern climates have asphalt shingles, which provide good traction on most of the roof, but at the edge of the roof the asphalt changes to sheet metal; this metal edge prevents the accumulation of snow. At night, unseen by a firefighter walking to the roof's edge in order to relay a message to the command post, this sheet-metal edge could be deadly.

Other Roof Hazards

Even if the size-up of the peaked roof reveals that the collapse danger, slope, and roof shingles are not great hazards, there are other dangers to a firefighter operating on the roof above a fire.

CLIMBING FROM A GROUND LADDER TO A ROOF LADDER

Firefighters have fallen and been injured when climbing from a ground ladder placed against the side of the building on to a hook ladder placed on top of a peaked roof. This sort of injury can occur when the ladder placed on the roof has an overhang extending beyond the eaves of the roof. If a firefighter, transferring his weight from the ground ladder to the roof ladder, applies downward pressure near this section of the overhang, he will lift the other end of the ladder off the roof ridge where it is secured. Both the firefighter and the roof ladder will fall to the ground.

To prevent this type of fall and to make the transfer from ladder to ladder as safe as possible, the ground ladder should extend four or five feet above the eaves of the roof, tools carried by the firefighter should be placed on the roof and hooked to the roof ladder, and the firefighter should hold on to the ground ladder for support as he steps on to the roof ladder. He should place his weight only on a rung of the roof ladder that is resting directly on the roof and avoid stepping on any part of the overhang of the roof ladder.

ROOF ROTTING

A deck below roof shingles may have rotted away from moisture accumulation over the years. Poor roof drainage causes a wood roof deck to rot. In many instances, the roof deck is not repaired—only a new shingle covering is placed over the rotted and decayed roof boards, concealing the danger.

If a firefighter walks on a section of roof that has largely rotted and decayed and a small area suddenly collapses, the firefighter may trip or fall down the sloping peaked roof and roll off the roof edge to the ground below. Firefighters must know the danger points on a sloping roof where deck rotting often occurs and avoid walking near them whenever possible. They are found near the edge of the roof, where the roof changes slope, and where a sloping roof abuts a vertical plane such as an additional story or a parapet separating row houses (Figure 7.13).

SKYLIGHTS

Flat, low-profile glass skylights are installed in many older private dwellings. During a smoky fire or at night in the dark, firefighters

Roof Rotting on Peaked Roofs

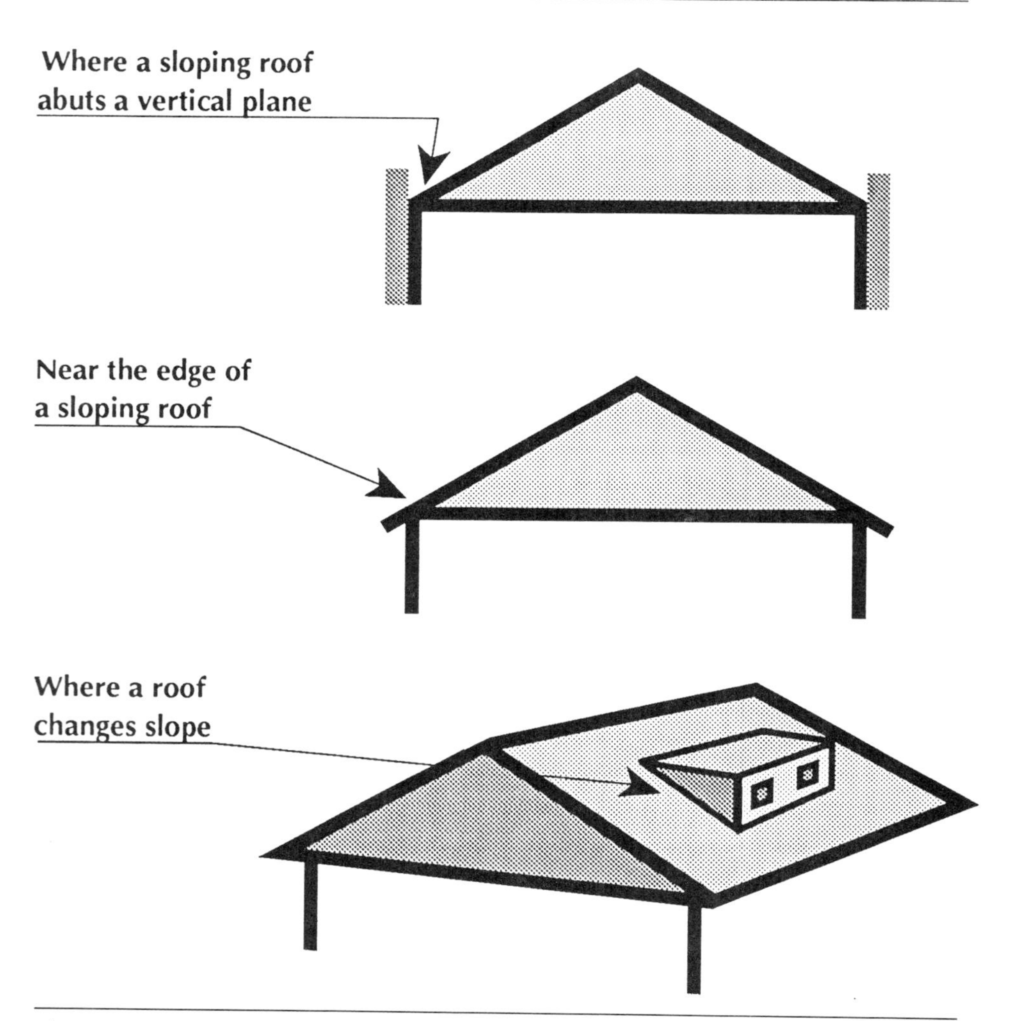

Figure 7.13. *A firefighter should know the areas subject to roof rotting from moisture.*

may not see these skylights, may step on them, and may crash through the glass opening and fall into a top-floor fire. When a skylight is installed, the roof beams are cut away, creating a large opening in the roof through which a firefighter could easily fall.

Walk near the roof ridge for protection—a skylight will normally be located several feet down from the ridge of a sloping roof. If your vision is obscured by smoke or darkness, crawl on your hands and knees along the ridge, feeling for shingles in front of you. Never walk blindly on the top of any roof.

SCUTTLE COVERS

Some peaked roofs have square or rectangular scuttle covers (hatch openings). They enclose access openings and ladders that provide admittance to the roof from the top floor of the house. Firefighters should be careful not to step on them; the cover could collapse and the firefighter fall through.

Scuttle covers are not designed to support the weight of a firefighter. Unlike the rest of the roof, there are no roof rafters below them. Treat them as you would a skylight. In fact, some scuttle covers are actually old skylights that were covered instead of repaired when they leaked or broke. Sometimes the material used to cover the skylight is merely roofing paper or thin wood and shingles. The owner may reason that, since the original skylight was not designed to support the weight of a person walking on the roof, the replacement cover does not need reinforcing. Stepping on a roof scuttle cover is an unsafe act that could lead to disaster.

DISORIENTATION

It is easier to become disoriented and lost on a smoky roof than in a smoke-filled room. "How could that be?" you may ask. The reason is that there are fewer guides on a large open roof to help firefighters reorient themselves. For example, on a roof there is no hose line to follow back to safety; there are no partition walls to touch or brush against in order to maintain a sense of direction.

On a smoky roof, unlike the smoky rooms below, the firefighter is often working alone (Figure 7.14). There are no voices of other firefighters to lead him in the direction of safety.

A disoriented firefighter on a peaked roof is exposed to many dangers: the roof edge, a skylight, a scuttle opening, a roof vent opening just cut in the roof, and, most of all, the slope of the roof. If you are on a peaked roof, momentarily blinded by smoke or darkness, and there is no immediate danger, remain in place and don't move. The wind that brings the enveloping cloud of smoke often changes direction and clears the smoke.

If there is danger and evacuation is necessary, get down on your hands and knees, carefully feeling the roof deck, and crawl up to the ridge of the roof. At this point, cross over to the other side of the ridge. The chances are that the smoke will not be severe on the other side of the sloping roof, and visibility will be better.

If the smoke still obscures your vision, straddle the roof ridge and crawl carefully to the edge of the roof to signal for help. When blinded by smoke on a peaked roof, never crawl downward on the sloping surface. Stay at the roof ridge, straddling the highest point.

Figure 7.14. *A firefighter can become disoriented by smoke on a peaked roof.*

Figure 7.15. When a fire department requires firefighters to operate on top of a peaked roof, the department should provide an aerial ladder for protection against collapse.

FIRE CUTTING OFF ESCAPE

When the fire size-up indicates that roof venting can be safely carried out on a large private house, place the portable ground ladder and roof ladder at the end of the roof opposite from that where cutting is to take place. A firefighter can then climb up to the roof ridge, where the roof is most stable, and walk to the other end of the roof along the ridge to the area above the fire.

You can make the roof cut for venting while straddling the ridge rafter. During operations on a peaked roof above a fire, however, you should remain continually aware of the path of escape back to the ladder. Never allow flames or smoke billowing out of a roof vent opening to cut off your escape route. Realize that, immediately after you cut a roof vent opening, it is possible for flame, heat, and smoke exploding out of the opening to block your path to the roof ladder and even to reduce your visibility so you cannot see the ladder. Thus, when venting a roof above a serious fire, just before completion of the vent opening position yourself between the escape ladder and the opening.

LADDER PLACEMENT

Standard operating procedure that requires firefighters to climb up on a peaked roof to extinguish, ventilate, or overhaul fires should also provide members with an aerial ladder or aerial platform in order to work safely (Figure 7.15). An aerial platform or ladder provides the safest and most efficient method of getting to the top of a peaked roof.

Figure 7.16. ABOVE: *Firefighters operating on the cornice of a peaked roof building.*
Figure 7.17. *A firefighter falling off the collapsing cornice.*

Lessons to Be Learned

Training firefighters in the safest techniques of walking and using ladders on peaked roofs will only protect them from some of the roof dangers, primarily falling and a local roof deck collapse—but it will not protect against a roof rafter collapse.

When the entire roof collapses, the rafters, roof deck, and shingles cave in to the top-floor fire (Figures 7.16 and 7.17). Any firefighter walking at the ridge or supported by a roof ladder will tumble into the top-floor fire. This type of peaked roof collapse has occurred, and an aerial ladder or aerial platform is then the only means of protection. A firefighter venting a roof while standing on the rungs of an aerial ladder or the basket floor of an aerial platform is independently supported by the truck parked in the street or yard below. The entire roof might suddenly collapse, but the aerial platform or aerial ladder will support the firefighter.

8

Cellar Fire Dangers

THE FIREFIGHTERS START forcible entry on an old glass door with a cylinder lock one foot above the doorknob. The doorknob is turned and the door pushed inward to make sure that it is locked. One firefighter shines a handlight on the lock while the other taps the K-tool over the face of the cylinder, inserts the adz end of a halligan tool into the slot of the K-tool, then forcefully pulls up the other end of the halligan tool. The movement tears the cylinder of the lock out of the wood frame, leaving the large glass pane unbroken. Quickly, a key tool is inserted into the hole and turned clockwise; the lock clicks, and the door is pushed into the store. Smoke flows out the upper half of the doorway.

Crouched down in the recessed store entrance behind the forcible-entry team is an engine company firefighter and an officer. Their face masks already donned, they bleed air from the hose line and replace their gloves and helmets. The hose team leads the way into a large, smoke-filled florist shop. The forcible-entry firefighters, now behind them, immediately spread out, searching for fire, victims, and vent openings. One firefighter remains at the doorway and feeds hose to the advancing hose attack team. The firefighter with the nozzle directs the hose stream at the ceiling heat and, while moving forward, knocks over potted floor plants and plastic flower pots stacked on top of one another. No flames are encountered. Visibility is near zero. The heat becomes more intense near the rear of the store.

As they advance in the smoke, the firefighter holding the nozzle lurches forward. The officer tries to grab him, but he, too, loses his balance and falls forward, then down, down, down. The officer lands on top of the firefighter. "What happened? Where are we?" The officer sees flames above at ceiling level. "We fell into the cellar."

Dazed and in a panic, the officer rips off his face mask and stands upright, but the heat and flames at ceiling level force him back down. Picking up his handlight and shining it upward, he sees that they have fallen through a cellar floor opening. A hinged "trap" door in the floor has been left open. The wooden stairway leading upward has burned away. They are trapped in the burning cellar. Cardboard flower boxes are burning near the front of the cellar. Flames fan out along the cellar ceiling and travel toward the opening through which they have fallen. Heat and smoke fill the entire cellar near the ceiling level.

"Help! Help! We are down in the cellar! Get us out of here!" shouts the captain. The firefighter, stunned by the fall on to a concrete cellar floor, rolls himself into a ball to avoid the heat. The captain jumps up and tries to grab the nozzle hanging down through the opening; the heat burns his face. He drops to the cellar floor. He picks up his handlight again and crawls to the rear of the cellar, away from the flames, searching for another way out. At the rear of the cellar, he wildly tears away piles of artificial flowers and plastic flower pots, hunting for an opening. He slides his hands along the solid concrete wall from one end to the other. No opening. Turning around and facing the fire at the front of the cellar with his back against the wall, he slides down upon a pile of straw on the cellar floor. He stares at the flames, grabs his portable radio, and calls: "Urgent! Urgent! Engine 8 is trapped in the cellar of the florist shop. Send help!"

Upstairs, on the first floor, a firefighter still searching in the smoke for the fire origin hears a distant call for help. It sounds as if it is coming from outside the store. He stumbles on the hose line on the floor. He crawls forward, following the hose line toward the nozzle. His flashlight beam is directed on the floor below the smoke in front of him. The heat becomes more intense.

Suddenly he comes upon a large opening in the wood floor from which smoke and heat pour out. The hose line is hanging down through the opening. He thinks: "We have a cellar fire, and the engine company fell into the cellar opening." Quickly he crawls out of the smoke-filled store. Once outside he yells: "Hey, firefighters are trapped in the burning cellar!"

The chief hurries to the firefighter and says: "I heard them on the radio. Where are they? Show me." The chief seizes another firefighter and orders: "Get a portable ladder into the store. Firefighters are trapped in the cellar." The chief and firefighters then enter the store. At the opening, the chief shines his light down through the three- by six-foot open trapdoor. Heat and brown smoke billow up through the floor opening, with small tongues of flame around its edges. The firefighter pulls the hose up out of the opening and then directs the hose stream down into the cellar.

The chief grips the transmitter on his portable radio. "Battalion One to Battalion One aide. Have a company vent the front store windows of the florist shop and have firefighters start breaking holes through the adjoining cellar walls to the florist shop. We have firefighters trapped in the cellar."

Just then a firefighter enters the store carrying a straight, 20-foot, aluminum ladder. "Where do you want it, chief?" "Over here, quick," replies the chief. "Slide one end down this opening."

Down in the burning cellar, the firefighter curled into a ball, with his knees pulled up to his chest, is struck by the water from the hose stream directed from above. Cool water splashes over him. He suddenly realizes someone is up there; someone knows they are trapped down in the cellar.

Rising up on one knee, he is struck on the shoulder by the ladder descending through the opening. Grabbing the ladder beam, he pulls it down to the cellar floor. Stepping on the bottom rung, he quickly begins to climb upward through the heat and smoke. Near the ceiling, the heat becomes intense. As his head pokes through the opening in the smoke, flame and heat rise up and burn his facepiece and neck. Quickly he ducks back down below the opening. The hose stream strikes him on the head and shoulders.

"There's one of them!" shouts the chief. "They're alive down there! Keep that hose stream operating." The trapped firefighter again tries to climb up the ladder through the smoke and heat. The pain becomes unbearable. He jumps off the ladder and falls to the cellar floor. He crawls beneath the shower of the cool hose stream. Soaking wet, he lies there. He feels the captain crawl over him. He watches him climb up the ladder.

The firefighter knows his SCBA will run out of air. "I do not want to die alone down in this cellar. I am getting out of here, too," he says to himself. He jumps on the ladder again, right behind the captain. At the middle of the ladder he feels the captain above him start to back down the rungs. Knowing the heat and pain the officer above him is experiencing, the firefighter nevertheless climbs upward. The officer starts to push him back down the rungs. The firefighter realizes that if he retreats, they will both die. The cellar flames are growing bigger.

In a frenzy, the firefighter places his shoulder beneath the captain and pushes upward, with all the strength left in his body. The captain pushes down hard; the firefighter is knocked down a rung. "Up! Up!" he screams into his facepiece. In a rage, the firefighter pushes upward, using his legs and shoulder. The captain bolts up through the opening. The firefighter stomps up the ladder behind him. As he rises up the ladder, the flames and heat again burn his head and wrists, then sear his face and neck. At the first floor, he leaps off the ladder to the side and rolls on the floor, his turnout gear smoldering. The hose stream is sprayed on him as he stumbles and is helped out of the burning store with the captain.

From a firefighting point of view, a cellar is the most dangerous area inside a building. More firefighters are killed and injured battling cellar fires than are killed and injured battling fires on any upper floor. In the past four decades, 30 FDNY firefighters died in cellars, killed by fire, explosions, collapse, drowning, and toxic gases.

Definition of a Cellar

Why are some cellars more dangerous than others? Firefighters should know which below-grade areas are most dangerous; some of these areas include crawl spaces, basements, cellars, and subcellars.

A crawl space is defined as a small space beneath the lower floor of a structure that allows access to wiring or plumbing. The crawl space area between the under side of the first floor and the unexcavated ground is not high enough for a person to stand upright.

A basement, on the other hand, is a full story below grade; one-half or more of its height is above grade level. When calculating the height of a building, the basement is counted as the first floor.

A cellar is different. It is a below-grade area that has more than half its height below grade. A subcellar is an underground level below a cellar. It is possible for a structure to have all three below-grade levels: a basement, a cellar, and a subcellar (Figure 8.1).

The degree of danger to a firefighter during a smoky fire is related to the depth below street level of the below-grade area. The deeper underground the cellar, the greater the danger during a fire. Access, extinguishment of flames, ventilation of smoke, and rescue will all be more difficult.

Some cellars are more dangerous than others. A cellar that is below grade at the front of the building but above grade at the sides and rear is not a true cellar and is not as dangerous as a cellar that is entirely below grade (Figure 8.2). The cellar of a building that has had the sides and rear yards excavated will have many windows and

Three Below-Grade Levels

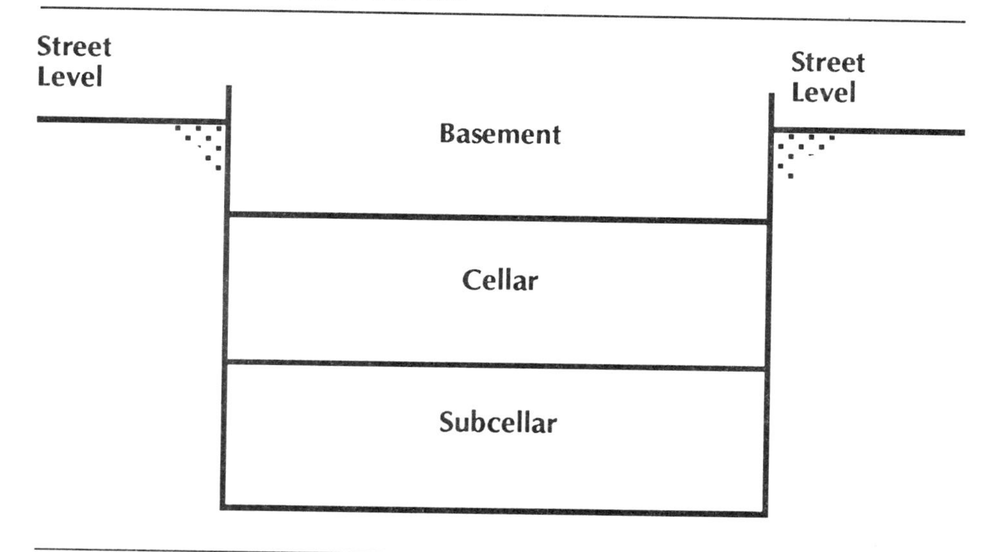

Figure 8.1. Three types of below-grade levels: a basement, a cellar, and a subcellar.

Above-Ground Cellar

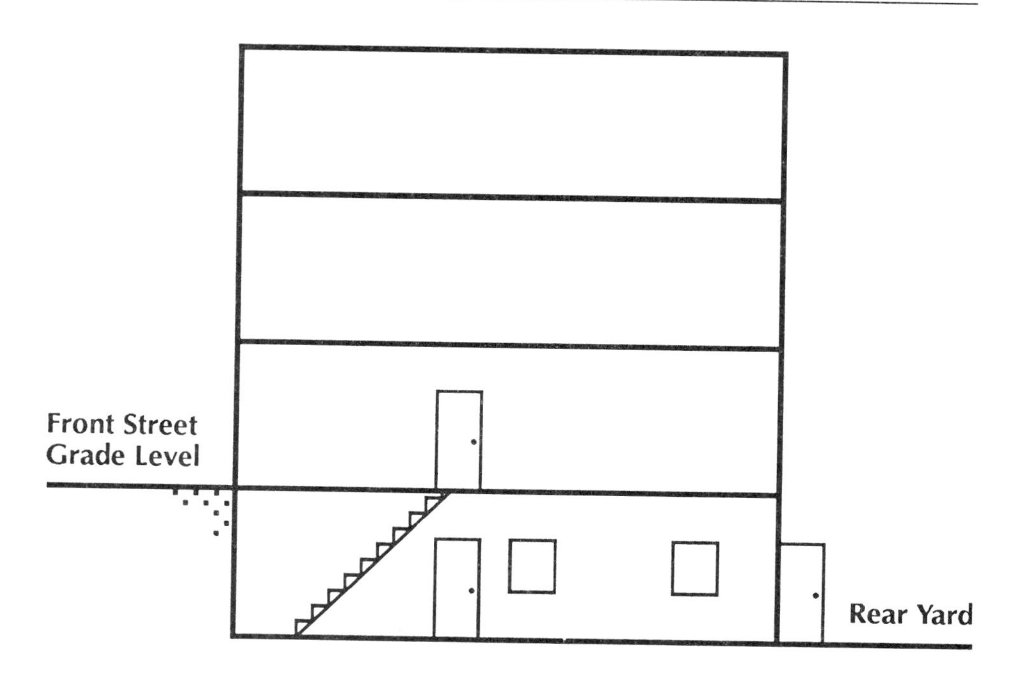

A partially above-grade cellar is safer to enter than one fully below grade.

Figure 8.2. Some cellars are above-grade at the rear of the building and below grade at the front of the building.

Below-Grade Cellar

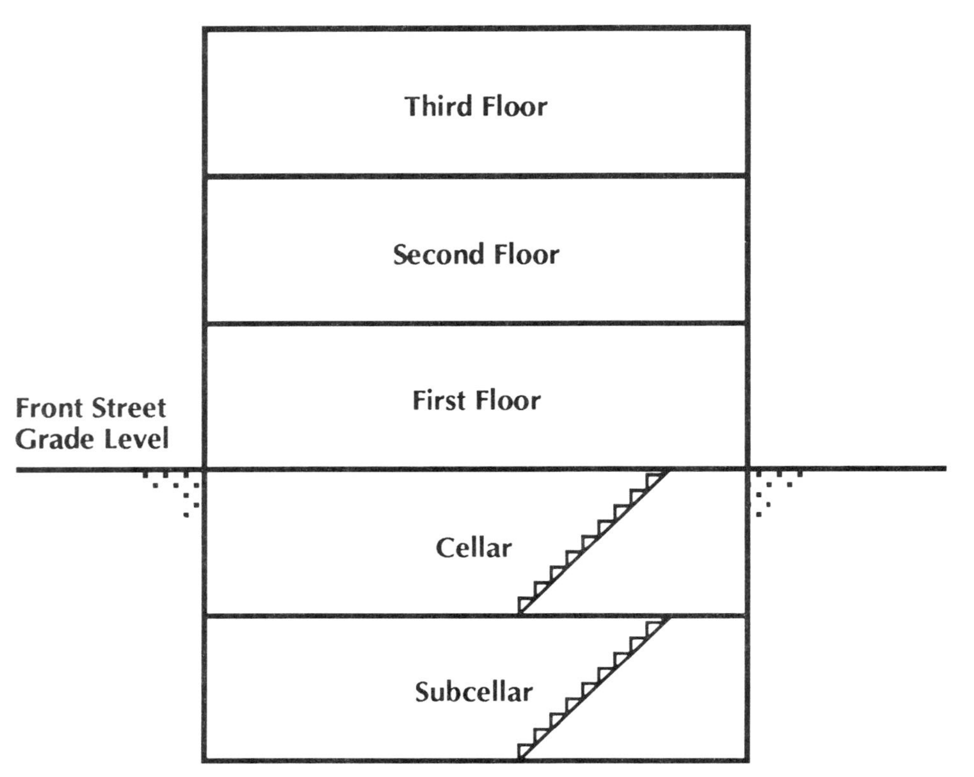

The absence of windows and limited number of entrances in below-ground cellars reduce ventilation and increase the dangers of carbon monoxide build-up and water accumulation.

Figure 8.3. *The most dangerous below-grade area is a cellar completely below street level, without windows for ventilation.*

several doors available to firefighters for access, ventilation, and rescue. By excavating the rear yard and designing side alleys around a building, 50 percent or more of a cellar can be above grade: because it is below grade at the front, it is still technically called a cellar.

A cellar that is totally below grade—front, sides, and rear—will have only one or two stairways and no windows (Figure 8.3). Because ventilation during a fire will be difficult to accomplish in this kind of cellar (and even when completed, it will be of a limited area only), deadly toxic gases can build up quickly, explosions of flammable gases can occur easily, and large quantities of water will be used for extinguishment due to limited access and smoke.

Only a subcellar presents more danger than a below-grade cellar. A subcellar possesses only one interior entrance and no windows; it is usually found in old commercial and storage buildings and in modern high-rise buildings. Fire in a subcellar is the most difficult below-grade fire extinguishment problem that can be encountered by the fire service.

Cellar Entrances

One of the dangerous jobs a firefighter must perform is to take a hose line down into a burning cellar stairway. Upon arrival of the first companies, the cellar stairway usually provides the only vent to the underground fire. Smoke, heat, and flame rise up out of the opening where firefighters must advance the hose line (Figure 8.4).

Figure 8.4. *Firefighters advancing an attack hose stream down a cellar stair must have great courage.*

In addition to having great courage, firefighters taking the hose down cellar stairs under these conditions must accomplish three dangerous maneuvers: First, they must drive back the flames lapping up out of the cellar opening while they are still at the top of the stairs. Then they must descend the stairs, through the heat and flame barrier rising up, and attempt to get below that heat and flame level in the cellar. Finally, the firefighters must pull the hose line across the smoke- and heat-filled cellar floor to the seat of the fire. To do this, they must often crawl through a maze of narrow aisles or climb over piles of paper or collapsed storage boxes.

The firefighters advancing the hose line down into a cellar could be trapped and killed immediately if the heat level has banked down to the cellar floor and there is no clear space below the flame and smoke. After several minutes in the cellar, if they become lost, disoriented, or if a concealed fire cannot be quickly found and extinguished, the flames and heat venting up the cellar stairs above the heads of the firefighters can completely fill the cellar stair—the only vent. If this situation occurs, the firefighters' only escape—back up the stairs—will be blocked, and they will be trapped in the burning cellar.

Cellar Stair Design

The design of a stairway leading down to a cellar determines the safety or danger to a firefighter entering a below-grade area. For example, a fire-retardant stairway enclosure, with a self-closing door at the cellar floor level, is the safest design for firefighting (Figure 8.5). Firefighters descending cellar stairs of this kind can take the hose line down to the cellar through a smoke-free stair enclosure, start water in the hose, check their masks and safety equipment, and then slowly open the door to the cellar. They are already below the level of heat and flame banked down from the cellar ceiling. If the heat is banked down to the floor and there is no survivable area below the flame, the door can be closed and the firefighters will still be safe in the stair enclosure. If there is too much fire in the cellar or the concealed fire cannot be located because of the smoke, but the heat and flames are increasing in intensity at the cellar level, the firefighters can be safely ordered out of the below-grade area. They can return to the safety of the stair enclosure. After closing the door at the cellar level, they can climb the stairs to the first floor in a smoke- and heat-free environment. They do not have to run through flame and heat rising up the cellar stairs. The door at the bottom of the stair enclosure acts as a protective barrier for the firefighters, keeping the cellar stair enclosure free of deadly flame, heat, and smoke.

Figure 8.5. ABOVE: *Heat, smoke, and flame rise up out of a cellar into the face masks of firefighters descending the stairs.*

Figure 8.6. *A sidewalk cellar entrance door leading to a below-grade area has steep concrete or wooden stairs, which make advancing an attack hose line dangerous.*

Unlike the enclosed stair, with a door at cellar level, the sidewalk cellar entrance stair is dangerous for firefighters. A pair of steel trapdoors, flush with the sidewalk when closed, located in the front or rear of a building, often covers a steep concrete or wooden stair leading down from the sidewalk to an open cellar (Figure 8.6). This type of cellar entrance has no fire-retarding enclose at the bottom of the stairs. Any flame and heat in the cellar will flow straight up out of the sidewalk cellar entrance, and there is a strong likelihood that the wooden stairs leading to the cellar have burned away or been weakened prior to the arrival of the firefighters.

Sudden collapse of a cellar step could cause a firefighter directing a hose stream to lose his balance and tumble into a burning cellar. Under ordinary conditions, the outside sidewalk cellar entrance stair is used more often to send boxes or cartons down to the cellar for storage than to permit people to enter the cellar. When this practice is common, the cellar stairs are covered with package chutes or rollers, designed to slide packages and boxes down over the steps leading to the cellar. These slippery chutes or roller ladders are sometimes left on top of the stairs after a delivery of stock has been stored in the cellar. During a fire, if firefighters attempt to descend the smoke-filled cellar stairway and step on a chute or roller, they will slide into the burning cellar. Escape from a below-grade area whose stairs are covered with a package chute or slide is unlikely. Before entering a smoky cellar, therefore, firefighters should check the stairwell and remove these dangerous chutes.

An even deadlier cellar opening is one type found inside a store. This opening leads to a cellar through a trapdoor in the floor. A wooden trapdoor, made of the same material as the floorboards, which is hinged at one side and opens on a wooden ladder leading to the cellar, is sometimes found in stores and storage buildings (Figure 8.7). Such a trapdoor creates a concealed three- by six-foot section of floor deck that is unsupported by floor beams below. Fire in a cellar will destroy such a section of unsupported floor deck first and cause what appears to be part of the floor to collapse into the burning cellar, taking any firefighters standing on it down into the cellar. These wooden trapdoors are usually located in an aisle or passageway so that stock or boxes will not cover access to the cellar. Firefighters, however, use these aisles to search or to advance a hose line when a fire occurs. Sometimes these trapdoors are left open at night when the store is closed. An unsuspecting firefighter searching a smoke-filled area could fall into the cellar through that open trapdoor.

Figure 8.7. Firefighters advancing an attack hose line can fall into an open wooden cellar trapdoor.

Elevators

In many modern buildings, there are no trapdoors or unenclosed cellar stairs leading to below-grade areas. There is, however, a new danger—elevators. Today, when firefighters respond to a report of a minor fire, building employees sometimes take them to the cellar or subcellar in elevators. Firefighters should realize the danger of this means of entry. If they are taken to the cellar by elevator to deal with an odor of smoke or a reported minor fire and the origin of the smoke or fire is found but proves to be a major blaze, they are trapped in the cellar with the fire. They do not know the location of the exit stair because they were taken below grade in the elevator. They cannot use the elevator to escape the fire because this shaft will turn into a chimney flue, venting the heat, flame, and smoke of the discovered fire. If the fire is beyond control of the hose line and the firefighters do not quickly find the cellar stairway exit back up to the street floor, they will die in the below-grade area. When firefighters are taken to a cellar by an elevator operator, they should determine the location of the cellar exit immediately, even before they rule the fire's origin.

Cellar Storage

If firefighters successfully descend hot, smoky cellar stairs and get below the heat barrier in the cellar, they must then advance the hose line through the cellar to the seat of the fire. Storage materials such as boxes, cartons, packages, and furniture can make the hose line advance difficult and dangerous. Large quantities of combustible merchandise or unused furnishings are frequently stored in cellars. This combustible material is stacked as high as possible, normally right up to the under side of the cellar ceiling. Also, there is usually only one narrow aisle through the stored material that leads to the utility supply and shut-offs. Such improper storage in a cellar creates dangers to firefighters advancing on attack line. It blocks the hose stream, preventing water from hitting the flames; it collapses and traps

Figure 8.8. *The reach of a hose stream allows firefighters to remain a safe distance from the flame, heat, and smoke of a cellar fire.*

firefighters in a burning cellar, and it conceals the exact point of fire origin, causing firefighters to remain for long periods in an oxygen-deficient cellar.

The safety factor provided by a hose stream reach of 30 or 50 feet allows firefighters to remain at a safe distance from the flames and heat of a fire (Figure 8.8). This advantage is eliminated in a cellar with storage material stacked up to the ceiling.

When a hose stream cannot be swept across the under side of a cellar ceiling because of piled-up storage, firefighters must crawl close to extinguish the blaze. A space of three feet or more between the top of the cellar storage material and the under side of the ceiling is necessary for effective use of a hose stream. Flames spreading through a cellar will be noticed quickly in that three-foot space, and the hose stream could drive the flames back, to a safe distance, if this three-foot space is maintained above cellar storage material.

Another problem caused by storage in a cellar is collapse of storage material. When cardboard boxes stacked on top of one another are burning or are wet by a hose stream, they will collapse. If firefighters advance a hose stream through a narrow aisle between cardboard cartons and the cartons collapse behind them, they may become disoriented or trapped in the cellar. The hose line, often used by firefighters to feel their way back to safety in a smoky cellar, is then covered with collapsed cartons of merchandise. Moreover, the narrow aisle, the path back to the cellar exit, is no longer present—just a pile of wet, burned, collapsed storage cartons.

Finally, another danger caused by cellar storage material is fire concealment. A fire in a cellar must be located before it can be extinguished. The exact point of origin of a fire may be hidden, smoldering behind boxes or furniture stored in a cellar. Firefighters may have to remain in the oxygen-deficient cellar atmosphere for an hour or more, searching for the concealed fire. Large amounts of storage may have to be removed from the cellar to reach the fire origin and completely extinguish the flames.

In some instances, a serious, rapidly spreading fire may be discovered behind storage material, piled up to the ceiling of a cellar. If there is no way to contain the flames temporarily and if the fire is beyond control of the hose stream on hand, there will be no time to wait for a backup hose line to be stretched to the cellar. The flames and heat will spread rapidly throughout the cellar, over the heads of the firefighters, and rise up from the cellar exit. The firefighters will have to abandon the hose stream and try to climb the cellar stairs before the flames and heat cut off their only escape (Figure 8.9).

Figure 8.9. Flames and heat rising up from a cellar entrance doorway can cut off the escape of firefighters down in the cellar.

Explosion

Confinement increases the chances of combustible gas explosion. A cellar is the most confined space in a building because there are fewer windows and doors leading to fresh air. The chance of an explosive mixture developing there when gas escapes is increased. Also, there are many combustible liquids and gases present that can form an explosive mixture. For example, a central heating system uses kerosene or fuel oil; a hot-water heater uses piped gas; flammable liquids are often found in a cellar.

During a cellar fire, a ruptured flammable-liquid container or broken gas pipe may release a flammable gas. If either is mixed with air and ignited by the fire, a violent explosion will occur. Firefighters may extinguish the cellar fire—and then be killed by the explosion.

The definition of an explosion is the rapid ignition of a combustible gas/air mixture that results in shock waves, structural collapse, and heat release. If neither shock waves nor structural collapse occurs, the phenomenon would be called a reflash or flash fire. Explosions in cellars during firefighting operations have blown street-level, plate-glass windows into the faces of firefighters standing on the sidewalk, blasted firefighters back up cellar stairways out on to the sidewalk, buried firefighters beneath collapsing masonry walls, and seriously burned firefighters near the explosion.

Recent studies reported in the NFPA *Fire Protection Handbook*, sixteenth edition (Quincy, Massachusetts, 1986), reveal some of the destructive effects of pressures created by an explosion:

Effect	*Peak Pressure Developed by Explosion*
Glass shattering	0–5 psi
Firefighter blown down	1 psi
Wood partition wall collapse	1–2 psi
Cinder block, non-bearing wall collapse (8 to 12 inches)	2–3 psi
Brick wall (non-bearing) collapse (8 to 12 inches)	7–8 psi

Armed with this limited information regarding the destructiveness of an explosion, a firefighter can reduce his chances of serious injury when a cellar explosion is anticipated by these precautions: *(1)* Stand clear of any windows that may suddenly explode outward. *(2)* Avoid standing near the entrance opening that will vent the pressure or shock wave of the cellar explosion. *(3)* Do not use a cinder block or masonry wall for protection against the force of the explosion. *(4)* Most important, wear all protective clothing—helmet, mask, gloves, boots, and turnout pants and coat. Even if a firefighter survives the explosion shock waves in a cellar, there will be a flash fire of extremely high temperature, created by the rapid ignition of combustible gas.

After a cellar fire is "knocked down," firefighters making a primary search often discover the melted connection of a gas meter, or a broken gas pipe burning. This finding should be reported immediately to the officer in command in the cellar, and the flame should not be extinguished. The flaming gas should be allowed to burn freely; the hose stream should be used to cool the surrounding combustible material. The gas supply to the broken meter or pipe should be shut off from a distant street or tank valve.

Even after the gas supply has been completely shut off, flame will continue to burn at the broken meter or pipe until the remaining gas in the system has dissipated. Firefighters should not extinguish this residual gas flame but allow it to burn out. Explosions have occurred in cellars from residual gas, even after the gas supply had been shut off.

Area of Room Needed to Contain Explosive Gas/Air Mixture for Explosion to Occur

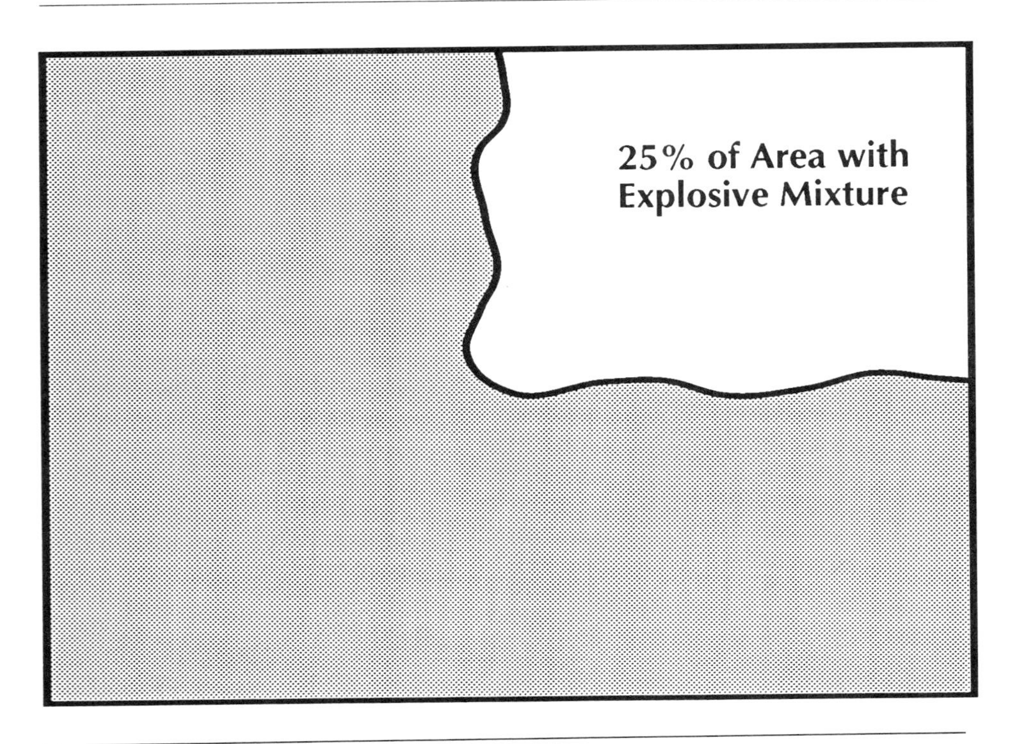

Figure 8.10. *A room in a cellar can explode when the combustible gas-air mixture is confined to less than 25 percent of a room.*

Another recent finding reported in the NFPA *Fire Protection Handbook* is that a flammable gas/air mixture confined to less than 25 percent of a room or cellar enclosure can cause an explosion (Figure 8.10). Based on this fact, it would be an error to assume that a flammable gas/air mixture must fill a cellar completely before an explosion can occur. An explosive mixture in one corner of a cellar could destroy an entire below-grade area, killing and injuring firefighters throughout the entire cellar, as well as those about to enter it.

Sometimes, after extinguishing a cellar fire, firefighters discover a broken gas pipe leaking gas that is not burning. This situation is more serious than a burning broken gas pipe, and fast action is necessary. The officer in the cellar should be notified, the hose line should extinguish any smoldering fire in the vicinity of the gas leak, the cellar area should be completely vented to prevent an explosive gas/air mixture from developing, and the gas pipe break should be plugged quickly with rags or soap while the supply is shut down.

Water Accumulation

If a cellar fire is not extinguished with the first attack hose line and high-expansion foam is not available, large amounts of water must be discharged into the cellar, using cellar pipes, distributors, and hose streams through the cellar entrance stairs.

In some instances, water may be flowed across the first floor above a cellar fire, according to Edward McAniff in *Strategic Concepts in Firefighting* (New York, Fire Engineering, 1974). Called "flowing the floor," it is used when the first floor becomes too dangerous for firefighters because of smoke or collapse. The objective of this tactic is to cover a porous floor in an old building with several inches of water in the hope that the water will drip down and enter the cellar ceiling and fire area. It is a last resort.

After any cellar fire has been extinguished, however, it is not uncommon to discover several feet of water in a cellar. Water accumulation is a danger. A firefighter collapsing in a water-filled cellar will drown. At an upper-floor fire, if a firefighter falls unconscious, he will be discovered quickly and removed to safety. If a firefighter falls unconscious in a water-filled cellar, he will sink out of sight below the water and drown.

There are many ways firefighters have been knocked unconscious and drowned in water-filled cellars. Firefighters have walked into low-hanging pipes or beams in cellars and struck their heads. Firefighters have been overcome by carbon monoxide, lack of oxygen, or oxygen depletion and collapsed. Firefighters have fallen into water-covered sumps or oil burner pits and struck their heads on objects there. Firefighters have been struck on the helmet by heavy sections of spalling concrete. Firefighters have fallen into cellars through holes in the first floor and struck their heads on the edges of the floor opening. Firefighters have fallen through collapsing terrazzo first floors and been knocked unconscious by the force of hitting the water-covered cellar floor.

Water accumulation in a cellar after a fire must be considered more than a nuisance that can indicate it is time to get new boots. Water accumulation is a deadly hazard of cellar firefighting.

In addition to drowning, water on the cellar floor after a fire can cause electrocution. If the water level has reached the electrical supply servicing the building and a firefighter enters the cellar, the electrically charged water will render the firefighter immobile and slowly shock him to death. Unable to move or to cry for help, the firefighter will collapse into the cellar water and either drown or die from cardiac arrest caused by electrocution. Electric and gas supplies should be shut off from the street before firefighters enter a water-filled cellar.

Asphyxiation

Firefighters are most often asphyxiated fighting cellar fires because their ability to vent a below-grade cellar is limited and sometimes impossible. In order to vent a cellar, firefighters may only have the option of being able to prop open one small cellar entrance or door or cut a small hole in the first floor directly above the fire with a power saw—but this effort takes time.

In many buildings, however, the first floor is constructed of brick, concrete, or terrazzo and can be penetrated only by air hammers. This task takes even more time. Before such venting can be accomplished, firefighters have usually been in the cellar for several minutes, either searching for the fire origin or directing hose streams on the blaze.

Asphyxiation or suffocation presents a great danger to firefighters in a cellar. Deadly carbon monoxide is most often stated as the major danger in a cellar death, but in fact asphyxiation can be caused by other products of combustion. For example, heavier-than-air toxic gases such as sulfur dioxide may have accumulated in the cellar. Furthermore, the oxygen in the cellar may have been replaced by carbon dioxide generated from complete combustion, or the oxygen may have been replaced or reduced to dangerous levels by steam, generated when the water stream struck the flames. If a firefighter dies from lack of oxygen in a cellar, it is a tragedy.

Self-contained breathing apparatus should always be worn before entering a cellar, regardless of the size of the flames or the amount of smoke present. SCBA should be worn during overhauling operations in an unvented cellar, as well, because asphyxiation deaths and injuries to firefighters in cellars occur most often at the beginning and the end of a fire operation (Figure 8.11).

Firefighters are not asphyxiated in cellars during actual hose line advance and fire-extinguishing operations. The following examples of firefighter deaths and injuries in cellars illustrate this statement best: Three firefighters were killed in a cellar when shutting off utilities after a fire had been extinguished. At another fire, a fire officer was pulled out of a cellar, unconscious; he had been almost asphyxiated by toxic gases while searching for the fire location in a light haze; a sprinkler had extinguished a small fire in plastic materials. At another

Figure 8.11. *Self-contained breathing apparatus must be worn before entering a cellar regardless of the size of the fire or the amount of smoke. Carbon monoxide is colorless and odorless.*

fire, two firefighters overhauling in a cellar became dizzy, staggered to the street, and collapsed at the command post.

Collapse of First Floor into a Burning Cellar

When unsuccessfully searching the first floor for the fire origin during a long period of time, suspect cellar fire. Continue to search the first floor, but send a firefighter down to the cellar to search for fire, too. The blaze could be below, sending smoke and heat up to the floors above. A cellar fire burning undetected for a prolonged period of time can destroy wood floor beams supporting the first floor, causing it to collapse into the cellar.

This hazard increases when the first floor is constructed of a terrazzo-finished cement laid on top of an old wood beam floor deck. Terrazzo floors consist of highly polished marble chips set in several inches of cement. This terrazzo floor increases the dead load supported by the floor deck and also conceals a cellar fire. Smoke and heat from the cellar fire will not rise up through a terrazzo floor as they will with a wood floor.

The cellar fire will first burn away the wood beams supporting the terrazzo-finished floor above them. There will be no sagging, springy, or spongy feel to the terrazzo floor to indicate potential collapse. After the wood beams have him burned away, the floor load is transferred to the terrazzo floor and then there will be a sudden collapse of that floor into the burning cellar. Any firefighter on that terrazzo floor will fall into the cellar.

Terrazzo floors often are found on the first floor of churches, restaurants, hallways, lobbies, bathrooms, kitchens, and stores. The presence of first-floor terrazzo flooring above a cellar fire should be considered a collapse danger. It is not possible to detect the weakening of a terrazzo floor by fire, nor is it possible to detect weakening of the wood support beams below just by inspecting them visually.

In some instances, however, a small amount of water from a hose stream sprayed across the terrazzo floor will evaporate quickly or turn to steam, which will indicate intense heat from the cellar fire below, conducting up through the cement and marble chips. On October 17, 1966, 12 FDNY chiefs, company officers, and firefighters died when a terrazzo floor they were standing on collapsed into a burning cellar fire.

Lessons to Be Learned

1. When a fire company enters a burning cellar, one firefighter with a portable radio should be stationed at the top of the stairs leading down to the cellar. This firefighter should warn the firefighters in the cellar to evacuate the below-grade area if heat and smoke rising up from the cellar increase in intensity to the point where they might prevent the escape of the firefighters.
2. Self-contained breathing apparatus should always be worn before entering a cellar even if there is only a light haze of smoke. Carbon monoxide and other toxic gases are colorless and odorless; they can be present in deadly quantities in a cellar even without smoke.
3. When entering a cellar to shut off utilities, wear self-contained breathing equipment.
4. Most incidents of firefighters being overcome in cellars occur after the fire is out and overhauling has begun. Toxic gases build up in a smoldering cellar during overhauling because

heat from the smoldering fire and hot concrete walls creates a positive pressure in the area. This heated, pressurized air expands and rises out of all the below-grade area vent openings and prevents fresh air and oxygen from entering the heated cellar through these openings. Two openings are required to ventilate a below-grade cellar: one opening to eliminate smoke, one opening to allow fresh air into the cellar. Electric fans should be used to force the ventilation (after the fire has been extinguished): one fan should be used to remove smoke from one cellar vent opening and the other to force fresh air into another cellar vent opening (Figure 8.12).

Figure 8.12. *Portable fans should be used to pump fresh air into a cellar and draft air out of a cellar as soon as the fire is under control, before overhauling begins.*

5. At several fires where firefighters have died or been overcome in cellars, there has been a light smoke condition and a sprinkler discharge controlling the fire. Do not let the presence of an operating sprinkler give you a false sense of security. Wear your SCBA.
6. Never enter a cellar without notifying your officer.
7. When a call for help is received for a firefighter "down" in a cellar, do not attempt a rescue without SCBA, or you will only add to the problem.

9

Propane Gas Fire Dangers

Two engine companies and two ladder companies respond to a report of a truck fire. The officer of the first pumper on the scene sees smoke issuing from a large, enclosed, lunch-wagon van. It's parked alongside a curb, at the spot where a row of two-story frame dwellings end and a school playground begins.

The lunch wagon was open for business when the fire started. A hinged metal panel on the street side, three feet high and five feet wide, is held open over the serving area by two metal poles. Flames float from the top of this opening.

"Pull up just past the truck," the captain of the engine company tells the driver. "We'll use the booster."

Speaking into the radio handset, he gives a preliminary status report. "Engine 8 to communications, we have a truck fire at King and Maple Streets. No additional units required. Return the rest of the assignment."

The officer then slides open the glass window behind him and briefs the three firefighters sitting in the jump seat. "Stretch the booster and take a hook and halligan tool for overhauling."

"Okay, Captain."

The pumper stops in the middle of the road. The firefighters mask up and stretch the booster hose. The captain moves to the burning van, and checks the cab for victims. There aren't any.

"Here comes the water!" one of the firefighters shouts as the line is bled of air.

The captain steps back from the truck, positions himself next to the firefighter holding the nozzle, and gives the command: "Hit it!"

The stream penetrates the bowels of the truck through the open service area, and the flames are quickly driven back into the truck. The firefighters move in closer.

Whoosh! Boom!

The explosion is sudden, violent. A ball of fire blows back out of the truck's side opening and the hinged panel slams shut. Flames reach out of the open cab door on the driver's side.

"Stretch the 1¾-inch pre-connect!" the captain orders the firefighter behind the man who's holding the booster line. The officer points to the firefighter holding the pike pole. "Open that hinge panel with the hook."

When the booster and the 1¾-inch hose lines are ready for another attack on the fire, the firefighter with the hook crouches down below the flames, holding the hinged panel open. The hose streams blast away at the flames and, again, drive them back into the van.

Boom! A fireball and a blast of air erupt out of the truck. Two firefighters are knocked down. The hinged panel is blown off the truck, into the air over the firefighters' heads, to the other side of the street.

Recovering quickly, the firefighters struggle to regain control of the open hose lines whipping around the street. The officer shouts back to the driver, "Take a hydrant!" He jumps up on to the back step, pulls off the nozzle with several folds of 2½-inch hose, and drops it into the street, then gets into the cab and calls for backup help on the radio. "Engine 8 to communications, send the full assignment to King and Maple Streets."

The captain jumps from the cab and checks the exposure behind the burning lunch wagon. A five-foot sidewalk separates the truck from the two-story frame dwellings. An old man and a child are standing at windows, looking down at the firefighters in the street. A group of children have gathered behind the schoolyard fence, attracted by the fire and the explosion. The top of the truck is burning. A large orange-and-white canvas umbrella affixed to the truck is engulfed in flames, and windswept pieces of charred canvas are blowing into the nearby dwelling.

"Captain, there's got to be some kind of gas leaking inside that truck," one of the firefighters reasons. The officer nods.

They both approach the burning truck, crawling below the flames that are shooting out of the open cab door. The captain is able to get a glimpse of the back of the lunch wagon. There, among piles of burning plastic coffee cups, a 20-pound propane cylinder is rolling around the floor, hissing and spewing flaming gas from its valve opening.

Whoosh! Boom! A small amount of accumulated gas at the top of the truck explodes. One of the hose streams temporarily extinguishes the burning gas.

"Hold it! Back off with that hose line—we have a propane cylinder in there!" the officer warns. He thinks to himself, "What am I supposed to do? What is the procedure for a burning propane cylinder?"

"Captain, propane is bad news. What should we do?"

Whoosh! Boom! There's another small explosion inside the truck. Looking up at the dwelling, the captain sees the old man staring down at him, and remembers what he must do.

"Bring that 2½-inch hose line over here."

He crouches at the open cab door and turns to the firefighter who's ready and waiting with the hose line, its nozzle closed.

"When I crawl in there, keep the fog stream directed at the top of the cylinder while I try to shut it off."

The firefighter nods. There is a moment of hesitation. Then, adjusting his gloves and mask, the captain crawls into the truck on his elbows and stomach, below the reach of the flames. He approaches the burning propane cylinder under protection of the fog stream. He grabs the collar of the moving cylinder and steadies it with one hand. With the other, he grasps the control wheel and slowly turns the valve. He lowers his head, staring down at the metal deck of the truck just inches from his face. His entire body shakes as he twists the valve clockwise. . . .

Bam!

He freezes.

"Captain, it's okay—a burning tire just blew!" The firefighter with the nozzle shouts, over the rumble of the spray stream striking the metal walls of the truck and over the hiss of the propane cylinder.

The cylinder valve finally meets resistance. It closes tightly, and the hissing stops. "That's it, Captain; you got it," the firefighter says.

Crawling back out of the truck beneath the fog stream, soaking wet, the officer takes off his helmet, walks over to the pumper, and sits down.

Sirens are sounding nearby—the other companies are pulling into the block. Someone taps the captain on the shoulder. He turns around. It's the battalion chief.

"Say, Captain, what's all this? Three hose lines stretched, calling for the full assignment—don't you think you're overreacting?"

"Chief," replies the captain, "you should have been here a couple of minutes ago."

Figure 9.1. Firefighters advancing attack hose lines on truck fires, house fires, and garage fires have been killed and injured by propane gas cylinder explosions.

Over the past two decades, there has been a dramatic increase in the use of small, portable cylinders of liquified petroleum gases at home and at the workplace. These 20-pound cylinders—generally propane or butane—are used for cooking, home repair torches, lighting, and heating and are often found inside garages, house cellars, auto repair shops, plumbing warehouses, and housing booths at convention centers and indoor festivals.

There has also been a dramatic increase in the number of firefighters killed and injured by explosions of small, portable propane cylinders (Figure 9.1).

Firefighters advancing initial attack hose lines on routine fires in trucks, house fires, garage fires, and trash fires have unknowingly stepped right into propane cylinder explosions. The deadly by-products of such explosions—fireballs, cylinders hurtling like projectiles, flying shrapnel, and shock waves—can seriously burn, maim, or kill a firefighter, blast him out of a window, or cause the collapse of a building in which he's standing.

Before firefighters are trained in tactics to control a fire near a portable propane cylinder, they should be aware of the reactions of a propane cylinder to fire and of the deadly by-products of BLEVEs. Knowledge of these reactions and explosive dangers is absolutely necessary for the fire officer in his risk assessment of fireground operations when propane cylinders are involved in a routine fire.

The Fireball

A BLEVE is not the fireball. A fireball is but one result of a BLEVE; it's caused by the instant vaporization and ignition of the liquid propane released to the atmosphere when the cylinder blows apart. Propane in liquid form expands 270 times in volume as it converts from a liquid to a vapor. One cubic foot of propane can thus expand into 270 cubic feet of vapor, with sudden explosive results.

The size of the fireball created by a propane cylinder BLEVE depends on the amount of liquid stored in the cylinder. These containers aren't completely filled with liquid; room is allowed for expansion of the liquid due to temperature variations. Before use, these common 20-pound cylinders hold about 80 percent propane liquid and 20 percent propane gas. Therefore, a so-called 20-pound cylinder will contain about 17 pounds of liquid propane, or approximately four gallons of liquid propane, weighing approximately 4.23 pounds per gallon. One gallon of propane liquid, if completely vaporized during a BLEVE, could create a fireball 36 cubic feet in volume; that's approximately three to four feet in diameter. A "full" tank—four gallons—of liquid propane vaporizing completely during a BLEVE could create a fireball of 144 cubic feet, approximately five to six feet in diameter. Of course, fireballs don't travel outward in neat, mathematical packages; the distance covered could, in reality, be considerably more than six feet.

Most firefighters have seen photographs of large storage tank explosions that produce gigantic fireballs several hundred feet in diameter. They're aware that death from burns has occurred to firefighters within a radius of 250 feet from large storage containers of liquefied petroleum gases.

Of course, the fireball created by a small propane cylinder explosion won't be nearly as large. Firefighters using the reach of a 30- to 50-foot hose stream on a 20-pound propane cylinder will be safely positioned out of range of the fireball, which may occur in the vicinity of the exploding vessel. The firefighters are, however, in great danger from rocketing cylinder parts, from a trail of burning liquid propane, and from the air blast of the explosion. These, even more than the fireball, are factors that endanger firefighters.

Projectiles and Shrapnel

Firefighters directing a hose stream could be decapitated or crushed to death if struck by a chunk of steel cylinder rocketing through the air at sonic speed. They could also be doused with burning liquid propane sprewed out in the wake of the rocketing cylinder.

It's impossible to determine exactly how a propane cylinder will break apart during a BLEVE. All we can say with certainty is that it will fail at its weakest point during a fire. That weakest point will be determined by the cylinder's exposure to fire—the heat and pressure weaken the molecular structure of the metal casing—or by any defects or damage to the container from handling.

In some propane cylinder fires, the control valve mechanism blows off and becomes a deadly piece of shrapnel, shooting outward far beyond the area of the fireball. At other fires, the tank splits at the seam and separates into two parts. If the tank is full of propane at the time of explosion, the rocketing cylinder will leave a trail of burning liquid and vaporized propane.

The distance that a piece of steel can be blown away from the explosion depends, in part, on the size of the container and the amount of liquid petroleum gas stored inside. BLEVEs of large tanks have blown metal pieces up half a mile away from the explosion. Firefighters who were 800 feet away from such a BLEVE have been killed by hurtling tank parts. Obviously, a small cylinder will not cause shrapnel to travel as far as will a large propane tank; however, firefighters directing a 30- or 50-foot hose stream on an exploding small cylinder are within the range of rocketing projectiles and could be killed or seriously injured (Figure 9.2).

The distance covered by metal shrapnel from an exploding propane cylinder is also dependent on which section of the cylinder fails. If the cylinder remains in one piece and only the control mechanism and valve blow off, that cylinder will travel farther than if the tank splits into two large sections. A small piece of rocketing cylinder such as a control handle mechanism is not unlike a bullet or cannon ball. If, on the other hand, the propane cylinder splits apart or tears open at the seam, the large chunks of metal may not travel as far away from the explosion site; however, this type of cylinder rupture creates a larger fireball.

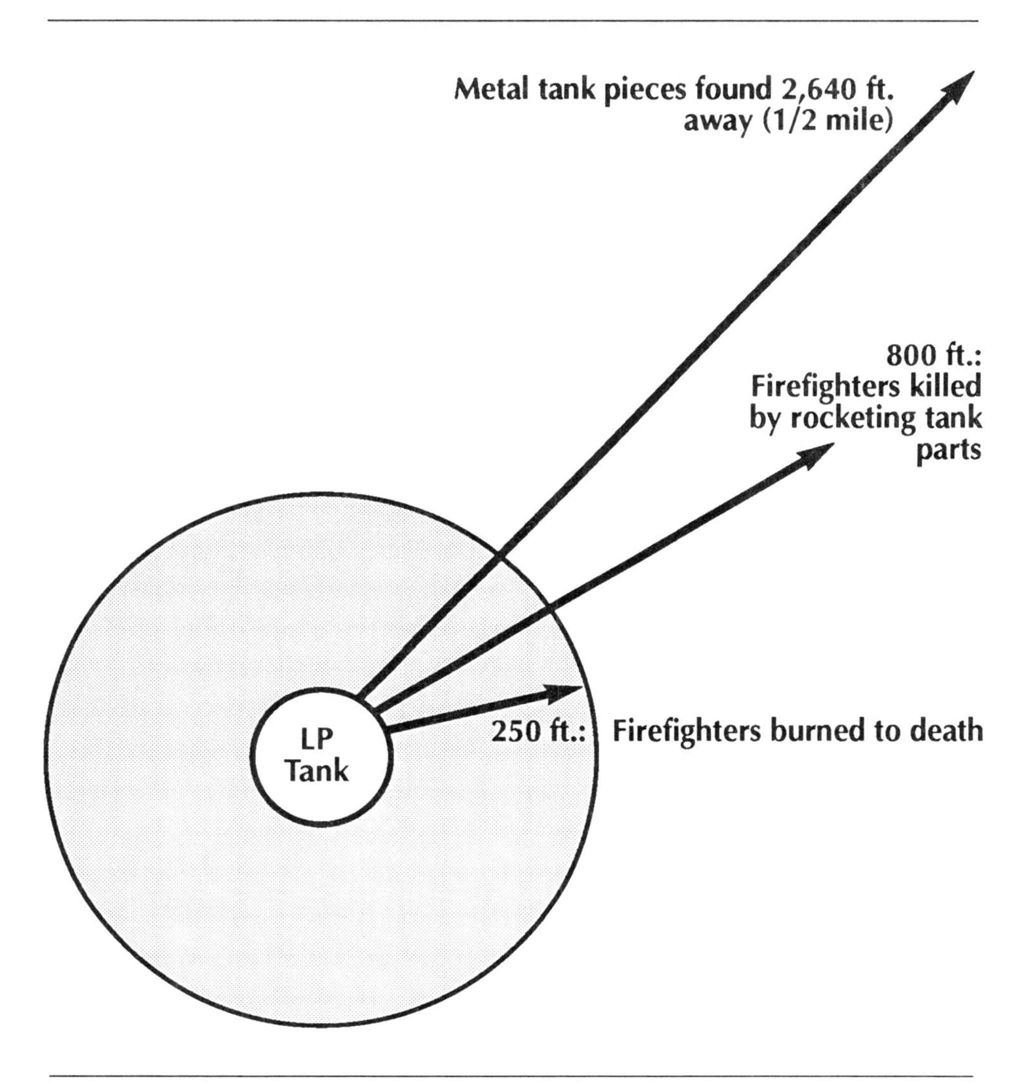

Figure 9.2. *Explosions of large propane storage containers have fatally burned firefighters within a radius of 250 feet.*

Test explosions and post-fire analysis have revealed that, when propane cylinders explode, the main section of the cylinder will rocket in a direction opposite that of the gas release or failure area of the cylinder. For example, if the bottom of a cylinder resting on its side "BLEVEs," the largest section of the container will travel in a direction away from the detached bottom of the cylinder. The area near the ruptured bottom of the tank will be the center of the fireball and the center of the greatest shock wave or blast.

Explosion Shock Waves

The definition of an explosion, according to the NFPA's *Handbook of Fire Protection,* sixteenth edition, is a "rapid release of high pressure gas into the environment."

This rapid release of high-pressure gas results in shock waves, which are actually blasts of air that transmit outward the high pressures created by the explosion. They may vary from supersonic shock waves to a mild wind. Their speed and force depend on how much "high pressure" is created by the explosion.

When an explosion occurs inside a building, the shock waves transmitted are confined by walls, floors, and partitions. These confined shock waves, if they don't cause the collapse of the building, can become deadly, concentrated blasts of air. Firefighters standing near openings during an explosion can be lifted off their feet and hurled into the air or out of a building. They've been blown off open floors of buildings under construction, blown into elevator shafts—together with the hoistway doors, and blown through glass windows by the shock waves of explosions.

Figure 9.3. *Explosion of a 20-pound propane gas cylinder can collapse a building.*

The deadly effects of shock waves from propane cylinder explosions can be extensive—far more extensive than the effects of fireballs and rocketing projectiles. When the shock waves create internal pressures that exceed the strength of the walls, roof, or floors of a building, massive total collapse occurs. On July 21, 1987, in Brooklyn, New York, one 20-pound propane cylinder exploded in the cellar of a plumbing supply company. The explosion and ensuing shock wave collapsed three two-story buildings, killing four people. Five years ago, in Buffalo, New York, the shock waves of an explosion caused by a leaking propane cylinder collapsed a 200- by 100-foot brick building, killing five firefighters (Figure 9.3).

When a propane cylinder inside a building is heated by fire and explodes, there are actually two explosions (and two shock waves). The first explosion is physical, which is the BLEVE or rupture of the propane cylinder; the second is chemical, which is the fireball or rapid ignition of the vaporized propane. The shock waves created by the physical explosion or cylinder rupture will travel outward from the side of the cylinder where the rupture occurs. The shock waves created by the second explosion or fireball may be more severe and extensive: these high-pressure shock waves travel in all directions.

The shock waves of the chemical explosion or fireball will be most destructive when they occur inside a building.

The explosion of a fireball and the subsequent collapse of a building will take place as follows:

1. The vaporized propane mixes with air.
2. The gas-air mixture is ignited.
3. Heat is absorbed by the surrounding air in the room.
4. The confined, heated air expands, doubling its original volume for every 459°F (237°C) that it has been heated.
5. The heated air, which is not free to expand (that is to say, heated air confined by the walls of the room), creates an increase in pressure.
6. The rapid increase in pressure creates shock waves that break windows, collapse interior partitions, and blow out the walls and floors of the entire building, if they are strong enough.

Vapor Space and Spot Fire

A 20-pound propane cylinder contains propane liquid and propane vapor: the liquid in the bottom area, the vapor, in the upper part. Most BLEVE failures of these cylinders originate in the metal of the vapor space (Figure 9.4).

The Vapor Space Is the Danger Area

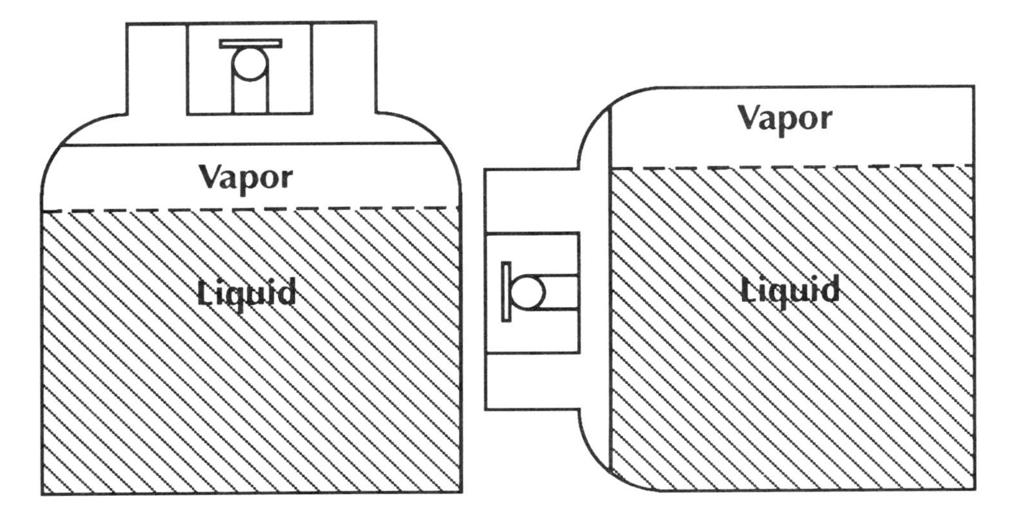

1. **Cool the vapor space of a heated cylinder.**
2. **Shut the gas off by the control valve if possible.**
3. **If the flow of burning gas cannot be shut off, allow the propane cylinder to burn itself out.**

Figure 9.4. *Hose streams should cool the vapor space of a propane gas cylinder.*

During a fire, it is extremely difficult to heat a cylinder to its explosion point when flame only contacts the liquid portion (or bottom). The most dangerous fire is one during which flame impinges on the upper portion of the cylinder, or the vapor space. Heat from the flames increases the temperature inside the container; the increased temperature causes the propane vapor to expand and weakens the metal shell of the cylinder; the increase of pressure inside the cylinder opens the pressure relief device; if a continued rise in pressure occurs, the cylinder will "BLEVE."

The highly dangerous condition of upper cylinder impingement is further intensified by a "spot fire," in which a propane cylinder is secured to a truck, a barbecue, or the side of a house, and flames issuing from a burning cylinder valve, relief valve, or tubing are deflected from a nearby object and back on to the metal enclosing the vapor space of the cylinder. Barbecue devices, truck bodies, and brick walls can deflect flames back on to the top of the cylinder.

Propane Cylinder Fire Tactics

When a large propane cylinder is burning around the cylinder valve, the correct firefighting tactics are: *(1)* cool the top vapor space of the container; *(2)* shut off the gas by means of the control valve if possible; *(3)* if the flow of burning gas cannot be shut off, allow the propane cylinder to burn itself out, and use the hose stream to protect exposures.

When a small, 20-pound cylinder is burning at the valve, it is often possible to shut off the flow of gas quickly without cooling it. A hose stream used to cool a burning 20-pound cylinder may accidentally extinguish the flame, and, if the gas reignites before the control valve can be shut off, a more serious condition could be created, injuring the firefighters attempting to shut off the valve.

The most common emergency involving a 20-pound propane cylinder leak is fire around the control valve or the tubing leading from the cylinder's control valve outlet.

Firefighters encountering this problem should first attempt to shut off the control valve handle. If the control valve handle is melted or missing, or if the valve is leaking gas around the packing nut below the valve handle, a vise grip or pliers may be used to turn the valve stem or tighten the packing nut. If metal tubing leading from the cylinder is leaking and the valve handle and stem cannot be manipulated to stop the leak, crimping or crushing the tube may stop the leak. Use a vise grip or pliers to "double-crimp" or bend the tubing. A "Z" type of crimp, similar to the one used on a burst hose line to stop, temporarily, the flow of water, is effective.

Fire officers supervising the hose stream at a propane cylinder fire must know safety principles of hose stream direction.

A fog stream at a 30-degree fog pattern will provide the greatest cooling effect and the greatest safety reach (Figure 9.5). The fog stream should always be directed at the top of the container or the vapor space in order to provide maximum cooling effect.

The fire officer must realize that a stream of water from a 1¾-inch hose line that's misdirected on a burning cylinder could extinguish the flame, and the escaping gas may then create an even greater danger from reignition—fireball, shock waves, and, if occurring indoors, collapse.

Safety demands that correct procedures be followed when shutting off the cylinder valve. The fog stream is adjusted from a 30-degree fog stream to a wide-angle fog stream (this change decreases the

Figure 9.5. *The fog stream should be adjusted from a 30-degree fog pattern to a wide-angle fog stream as you advance on a propane gas cylinder. This precaution decreases the likelihood of accidental extinguishment of the flame and possible explosion.*

Figure 9.6. BELOW: *Two fog streams angled in overlapping fashion may be used for maximum firefighter protection.*

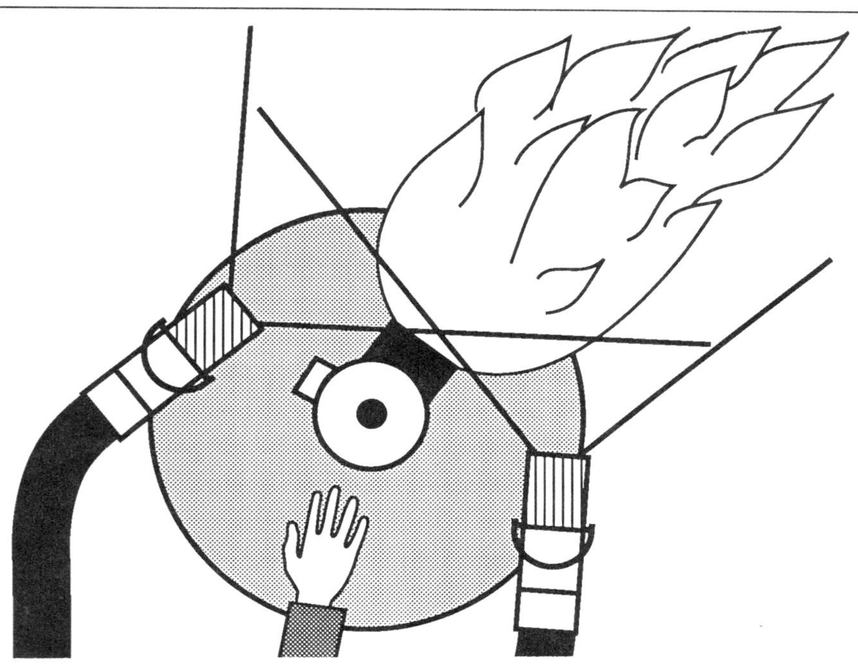

A firefighter shutting off the flow of burning gas at a propane cylinder outlet should be protected by a wide-pattern low-velocity stream; position the flow stream between the control valve and the burning outlet. The firefighter's hand should be behind the fog curtain when turning the control valve. The flaming outlet should be in front of the fog curtain.

likelihood of extinguishment). The fog stream is positioned between the shut-off valve and the burning outlet. Two fog streams angled in overlapping fashion may be used for maximum firefighter protection (Figure 9.6). Never extinguish a propane gas fire unless the fuel can be shut off at the valve immediately.

A recent propane cylinder incident brought about this course of events: After extinguishing a house fire, firefighters heard a loud hissing noise and discovered a scorched 20-pound propane cylinder among a pile of burned rubble. Its relief valve was discharging propane gas. A fire officer quickly picked up the discharging cylinder and threw it out of a window into the backyard. It safely discharged its contents until the tank had been cooled and the relief valve reset. Afterward, the officer agreed that such rough handling of the propane cylinder could have resulted in dislodging the control valve or denting the tank, creating a larger, unstoppable leak. He concluded, however, that this risk was preferable to the continued gas accumulation and potential explosion inside the structure.

A room can explode when less than 25 percent of the enclosure contains a flammable gas-air mixture. It's erroneous—and dangerous—to assume that a gas-air mixture needs to fill an entire room before an explosion can occur. Therefore, prompt removal of the leaking cylinder is imperative.

In some communities, owners of high-rise apartment houses prohibit tenants from keeping propane-fired cooking stoves on open-air balconies. Some tenants, naturally, try to circumvent the regulation by stowing the barbecue inside the apartment and sneaking it out on to the balcony for brief periods of use. Keeping the cylinder inside the apartment is not too clever, when the question of fire arises.

If, after a fire in such a building, a scorched cylinder is discovered with its relief valve discharging gas, it should be quickly carried out to the balcony and cooled with a 30-degree fog pattern stream. The windows and doors leading from the outdoor balcony to the apartment should be closed in order to prevent wind from blowing gas back into the apartment. If the relief valve is a resetting pressure device, it will close when the cylinder is cooled. If the relief valve is not of the resetting type, the cylinder valve should be opened to speed up the gas discharge and a hose fog stream played on the leaking gas to dissipate it into the open air.

Moving a propane cylinder requires that the following safety precautions be observed whenever possible:

- Firefighters should wear full protective clothing on all parts of their bodies: helmet, turnout coat, gloves, boots or bunker pants, nomex hood, and the donned face mask of the SCBA.
- The control valve of the cylinder should first be shut off.
- If there is a nearby fire exposing the cylinder to heat, the firefighter moving the cylinder should work under cover of a hose stream, or else a hose stream should be charged and ready for immediate use.
- The cylinder should be approached from the upwind side.
- The firefighter should carry the cylinder with its relief valve pointed away from his body; in that way, if the relief valve were suddenly to operate, the firefighter would not be drenched with propane gas or burned by a fireball.
- The wind should be at the firefighter's back during the entire process of moving the cylinder.

Figure 9.7. *The margin for error in firefighting tactics at a propane gas cylinder fire or leak is very small. Mistakes quickly turn into fatalities.*

- The 20-pound propane cylinder should only be moved in an upright position. If it's resting on its side, it should be set upright first, before moving. The relief valve is designed to operate when the vapor space is at the top of the cylinder. If the propane cylinder were on its side and the relief valve suddenly became activated, liquid propane could be discharged from the opening.

Lessons to Be Learned

Fires that involve burning or leaking 20-pound portable propane cylinders can be routine—or they can produce deadly explosions. If improper firefighting tactics are used in the handling of these cylinders, a boiling-liquid, expanding-vapor explosion could occur, a fireball could ignite around the cylinder, the metal tank could rocket 50 to 100 feet through the area at sonic speeds, a structural collapse could occur—any one of which might kill a bystander or a firefighter (Figure 9.7).

The margin for error in firefighting tactics at a propane cylinder fire or leak is very small; mistakes can quickly turn into fatalities. Fire officers must be thoroughly knowledgeable in this area in order to make rapid risk assessments and initiate proper strategy and tactics; firefighters must have the procedural knowledge to carry these out without hesitation. Safety demands it.

10

Wildfire Dangers

"BATTALION 1 TO COMMUNICATIONS. We have a brush fire traveling up the side of Blackstone Mountain. There appears to be a house near the top of the mountain that may be threatened by the fire. Send an engine company to check it out. There are no other exposures at this time. Communications Center, ten-four."

"Communications Center to Engine 8. You are being redirected from the main fire to check out an exposed residence building near the top of Blackstone Mountain. Our computerized information dispatch system shows a plot plan; take Scenic Road 500 yards past Route One and make a left turn. A property road extends from Scenic Road to the house. Engine 8, ten-four."

The captain of Engine 8 looks at the driver next to him and says, "I think I know the place. Down the road we should hit a sharp turn; there will be a mailbox; we make a left on to a dirt road and head down that road about 100 yards to the house."

Arriving at the mailbox, the driver turns the pumper left and starts down a steep, bumpy road, almost concealed by overgrown trees. The side view mirrors of the slow-moving engine bend and break tree branches and shrubs as the apparatus bounces and lurches from side to side down the rocky road. A light haze of smoke drifts in the air beneath the overhanging tree branches.

"Captain," says the driver, "we are heading in a bad direction. We are uphill of a spreading brush fire." "I know," replies the captain, "but we have to check out this house." They ride ahead in silence.

"Captain, look over there," the driver points to his right. In the woods, sparks and smoldering scraps of vegetation are floating down from the sky among the tree trunks. The radio interrupts: "Communication Center to Engine 8. We just received a telephone call from the occupant of the house. She can see flames coming up the side of the mountain, and she cannot start her truck."

"Engine 8, ten-four. We are almost there." The pumper rocks violently to one side as a wheel drops into a deep crevice of the dirt road. As the pumper slowly continues, it passes several small fires burning in the undergrowth. The woods on either side of the road are now filled with thick smoke. As the pumper moves forward, burning leaves and tree branches become caught in the windshield wiper blades of the front windows, the brackets of the side view mirrors, and the portable deluge nozzle on top of the rig; the pumper is virtually camouflaged with foliage. The driver reaches out his side window and pulls a small, broken bit of tree branch from the windshield. As he fingers the twig, he wonders if they will be able to get out of these woods after they have reached the trapped woman.

On this hot autumn afternoon, the woods are very dry. The forward-moving pumper pushes aside a clump of low-hanging tree limbs and finally enters a grassy area. Smoke fills the air. Ahead is a small house, almost concealed by shrubs and bushes. Near the front door, a pickup truck is loaded with furniture. A large, gray-haired woman steps out of the truck with a handkerchief to her face. "Thank God you came; my truck won't start and the flames are coming up the mountain."

The officer quickly walks around the house to size up the situation. He looks down the mountain and sees a large fire-front blazing up toward the house. Shrubs and small trees are exploding into flame. Heat waves and smoke roll up the mountainside ahead of the main fire (Figure 10.1). The captain turns to the house. The walls are brick, but the roof is covered with combustible wood shingles. There are no attic windows, but the ranch-type, low-sloped roof has a large overhang at the eaves, where the roof extends beyond the outer walls; this type of overhang will trap heat and flames. The shrubs are very close to the house. The captain must decide whether to defend the house or try to make it back through the woods with the woman.

Running to the front of the house, the officer shouts, "We are going to defend the place." Speaking to the driver and two firefighters standing near the woman, the captain orders: "Position the pumper at the center of the clearing, away from the trees. Throw a canvas tarpaulin over the pump controls. Engage the engine into pumps and charge the hose line."

Figure 10.1. Each year, firefighters are killed battling residence fires, wildfires, and store fires, in that order.

Moving toward the two other firefighters, he directs: "Place the ground ladder to the roof of the house, stretch the 1¾-inch line into the house, and stretch enough hose to protect the entire outside of the house. Bring the backpack water extinguisher inside."

He turns to the woman and says, "Go inside. We are going to let the fire burn over us, and try to save your house. Please hurry; we have only about five minutes before the fire reaches us."

After all the outside tasks have been completed and the firefighters assembled inside the house, the officer continues to direct the operation. Speaking to the firefighters, he points to one: "Go around the house and close all the windows, lower any shades, and close any shutters or curtains. Keep the fire from coming through a window." To another firefighter, he says: "Stay with the nozzle and close the front door as much as possible with the hose extending through the doorway." To the remaining firefighter, he directs: "Get that pump tank, go upstairs in the attic, and spray water around the eaves. Wet the attic floor where the roof slope ends—I want water dripping from the eaves outside at the roof overhang. Refill the tank and stand by near the attic opening. If flames come through the roof, wet them down with the water stream. Here, take my handlight up with you." Turning to the firefighter with the nozzle, he says: "After the fire passes, we go out that door with the hose line, climb up the ladder, and extinguish the fire burning on the wood-shingle rooftop."

"Captain!" a voice from a distance calls. "All the windows are secured." Another voice is heard: "Captain, all set in the attic. It's wet down." "Okay, good," the officer shouts back.

Suddenly, the oncoming fire can be heard outside. *"Thump, thump."* Airborne pieces of burning tree branches are falling on the roof. *Thump, thump, thump, thump,* they start to hit the roof with increasing frequency.

Then a strong, hot wind begins to blow around the outside of the house. Shutters bang back and forth against the house. The howl of the rising hot convection currents traveling up the burning mountainside ahead of the main fire sounds like a windstorm. Finally, the flames arrive. A loud roar engulfs the house. Those inside hear the shrubs around it burning: the dried-out leaves and twigs crack and pop as they burst into flame. Smoke begins to drift inside, through the partly opened front door and through the closed windows around their wooden frames.

One of the firefighters calls out: "Captain, fire ignited one of the window curtains." The woman is running through the hallway from the kitchen with a large pot of water for the firefighter. The sounds of the fire subside. Now it is quiet outside. "Are you all set?" the officer asks the firefighter with the hose line in front of him. "Yes, Captain," the firefighter responds. "How about you?" the officer asks the pump operator behind him. "All set," that firefighter responds. "Let's go." says the captain, as he pushes open the front door. The area around the house resembles a battlefield. The ground is scorched and littered with black ash. The bushes around the house have burned to the ground. The treetops of the woods beyond the clearing have been burned away; only charred branches and trunks remain. The furniture in the pickup truck is ablaze. The wood-shingle roof of the house is flaming.

While the officer and the firefighters quickly climb the ladder to the blazing roof, the pump operator runs over to the pumper, pulls the cover off the pump panel, and examines the control panel (Figure

Figure 10.2. *Priorities of fighting wildfires are: protect life first, protect property second, and contain the fire third.*

Figure 10.3. RIGHT: *Firefighters are most often killed or injured fighting small brush fires in isolated portions of large fires.*

10.2). Everything appears to be in working order. He slowly pulls out the pressure knob and increases the nozzle pressure of the hose. He surveys the vehicle; all the red plastic lights and front trim have melted from the pumper. Its glass windows are discolored and cracked from the heat. He looks back at the house. The captain and the firefighters on the roof are extinguishing the last of the roof deck fire. The firefighter with the backpack tank is wetting down the underside of the eaves with the stream of water. The woman stands outside the doorway, looking at her house.

Wildfire is a new name given to an old firefighting danger—brush and forest fires. Brush and forest firefighting is the second most dangerous activity in which a firefighter can engage. Each year, most firefighters are killed battling residence building fires, brush and forest fires, and store fires, in that order. In 1986, 17 percent of the fireground deaths took place at brush and forest fires.

There are many misconceptions about wildfires, which, each year, consume more than five million acres of land in the United States and Canada and leave 50,000 people homeless. The most serious misconception about wildfires is that they are not dangerous to firefighters. The second serious misconception about wildfires is that only large forest fires kill and injure firefighters. The actual facts are: Wildfires are extremely dangerous, and most firefighters are killed and injured by brush fires (Figure 10.3), not forest fires, according to Carl C. Wilson in his book, *Fatal and Near Fatal Forest Fires: The Common Denominator.*

A survey of fires revealed that firefighters are most often killed or injured at small brush fires in isolated portions of large fires. They are not killed by large timberland forest fires. The treetop or crown fire may destroy more acres of woodland each year and receive the greatest media attention; however, the brush fire kills firefighters. Firefighters are burned to death trying to outrun brush fires, or they are engulfed in flames when a brush fire suddenly flares up around them.

Burning bushes, shrubs, hedges, thickets, undergrowth, grain fields, and marshland grass are considered brush fires. These types of vegetation, called brush, grow from five to 20 feet high, are clumped together in thick masses, and are considered lightweight fuels. Trees, logs, and wood branches are considered heavy fuels. The light fuels of a brush area are the first to dry out during hot weather. Their finely divided leaves and twig ends are easily ignited. Because of the thick clumps of growth common in brushy vegetation, flames spread with explosive speed. In the case of pine and evergreen shrubs, a natural oil on the leaves adds fuel to the brush fire.

Dried-out, burning brush is extremely susceptible to changes of wind and changes in the slope of the ground. Flames of burning brush can reverse direction instantly when the wind changes. Flames of burning brush can suddenly race up the side of a hill or mountain. Flames of burning brush may flare up explosively or blow upward 20 feet into the air when fanned by a gust of wind. Firefighters near a brush fire are constantly in danger of being engulfed by a flare-up and burned to death (Figure 10.4).

Flare-Up

Flare-up and blow-up are terms used to describe how brush fires burn. Open areas of brush do not burn evenly. Flames do not progress from one side to the other of an open field or travel up a mountainside in one steady burn. These flames spread in spurts. Fire flares up in

Figure 10.4. *Flare-up can be caused by a gust of wind or a change of wind direction. A small ground fire can suddenly erupt into a raging inferno of flaming brush.*

one area and then dies down. It may flare up in another distant area, then die down again. It may suddenly race up the side of a mountain. Brush fire spreads erratically and unpredictably.

A gust of wind often changes a small ground fire into a sudden, ranging inferno of flaming brush: a flare-up. Firefighters trying to make their way through thick brush could be caught in a flare-up and burned to death. Fires during which firefighters are trapped and burned in flare-ups often appear innocent just before the flash. In some instances, firefighters have been killed by flare-up in the mop-up stage of the fire.

Types of Wildfire

There are three classifications of wildfires: ground fires, surface fires, and crown fires. The ground fire, sometimes called a bog fire, is a

Figure 10.5. ABOVE: *A ground fire, sometimes called a bog fire, is a slow-spreading, smoldering fire in dried-out leaves and twigs on the ground.*

Figure 10.6. BELOW LEFT: *Brush fire is a fast-moving fire involving grain fields, marsh weed, chaparral, scrub oak, or grass.*

Figure 10.7. RIGHT: *A crown fire is a treetop fire; it spreads from treetop to treetop.*

slow-spreading, smoldering fire, burning in dried, decomposed leaves and twigs on the ground (Figure 10.5).

The surface fire is the brush fire, a fast-moving fire, burning in grass, grain fields, scrub oak, hemlock, pine, chaparral, or marsh weed (cattails) (Figure 10.6).

The crown fire, sometimes called a forest fire, is a treetop fire (Figure 10.7). A crown fire spreads flame from treetop to treetop. Most crown fires are caused by vertically spreading brush fires.

Danger Areas and Brush Fire Spread

To survive fighting wildland fires, a firefighter must know the way a typical brush fire spreads. In the fire service, knowledge is safety; the knowledgeable firefighter is the safer, more effective firefighter. Brush fires spreading across a field create a recurring burn pattern. Aircraft pilots have recorded it, and the fire service has designated terms for areas of this burn pattern. A firefighter's safety and survival depend upon knowledge of the brush fire's danger areas and the significance of fire spread in each of these danger areas. The areas are named as follows (Figure 10.8):

- *Head:* The edge along which the fire is advancing most rapidly
- *Fingers or Points:* Long, narrow areas burning in advance of the head
- *Spots:* Areas of brush that have ignited in advance of the fire, caused by flying embers or radiated heat
- *Flanks:* The sides of the fire between the head and the rear

Burn Patterns of Wildfires

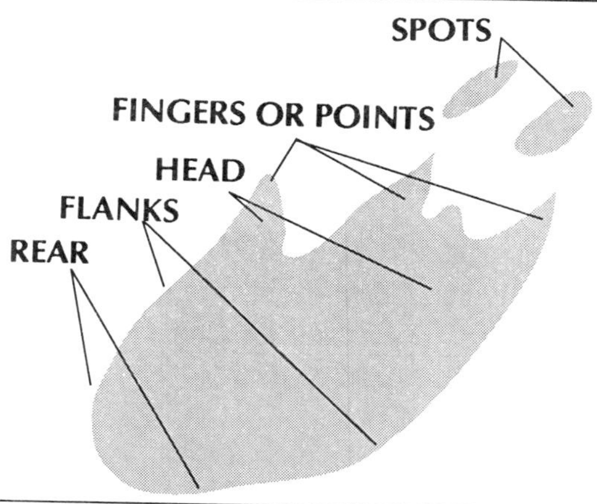

Figure 10.8. *The head of a brush fire is the most dangerous area.*

Figure 10.9. *Firefighters on a hillside being outflanked by flames spreading up the side of a mountain.*

As a brush fire moves across a field, the outer perimeter of the fire progresses unevenly. Long, narrow strips of flame (fingers) extend forward from the main fire area. This uneven fire spread is caused by wind variations and different rates of flame spread in the brush. A firefighter operating alone with a backpack extinguisher at the head of a brush fire can be outflanked quickly by the spreading fingers of flame (Figure 10.9). Also, spot fires started by burning sparks and embers, swept by winds over the heads of firefighters, could encircle and cut off the escape of a fire company.

The head of a brush fire is the most dangerous area; firefighters are killed or injured seriously at this location.

The flanks of a brush fire present less risk to a firefighter. Flame spread at a flank is not as rapid as that at the head of the fire; however, a sudden shift of wind direction can change one of the flanks into the head of the blaze. Caution should be used when operating on a flank area. The rear, or downwind, area of a brush fire usually burns less intensely and less rapidly then the head or the flanks and is a safe area in which to start a perimeter attack around both flanks of the fire. It is also a possible area suitable for setting up a command post or rehabilitation area.

The Blackened Area

The safest position for a firefighter to attack any brush fire is inside the burn area itself. All the combustible brush has already been consumed by fire. A safe, close approach to any part of the fire can be obtained from this area. If the wind suddenly changes, there is no fuel to burn; there is no danger of a firefighter being caught in a flare-up or blow-up.

There are, however, less serious hazards that a firefighter should be aware of when operating inside the burned-out area of a brush fire: Smoke from smoldering ground fire, tree trunks, and leaves permeates the area; the scorched fireground area will be obscured by a thick haze of smoke. Carbon monoxide is given off by the smoldering fires, and masks should be worn.

This haze of smoke will also reduce visibility and increase the danger of tripping and falling. Heat from the charred earth inside the blackened, burned-out area of the brush fire increases the danger of heat exhaustion. Firefighters' operating time, before exhaustion sets in, is reduced when operating inside a hot, burned-out area. Rehabilitation and rest areas must be set up where firefighters who have worked in a burned area can safely remove fire clothing, replenish body fluids, and cool off.

A Defensive Operation in Front of an Oncoming Brush Fire

Control of a brush fire at or near its head is the key to most firefighting operations. A quick attack on the head area is the strategic objective of most brush firefighting (Figure 10.10). The head of a fire can be approached safely from the burned-out area, or, sometimes, it can be approached from one or both flanks and the fire cut off. In some instances, however, control of the fire at the head can be accomplished only by positioning pumpers or brush vehicles at a road or house in front of an onrushing fire. This strategy is extremely dangerous and

Two-Pronged Flank Attack on a Brush Fire

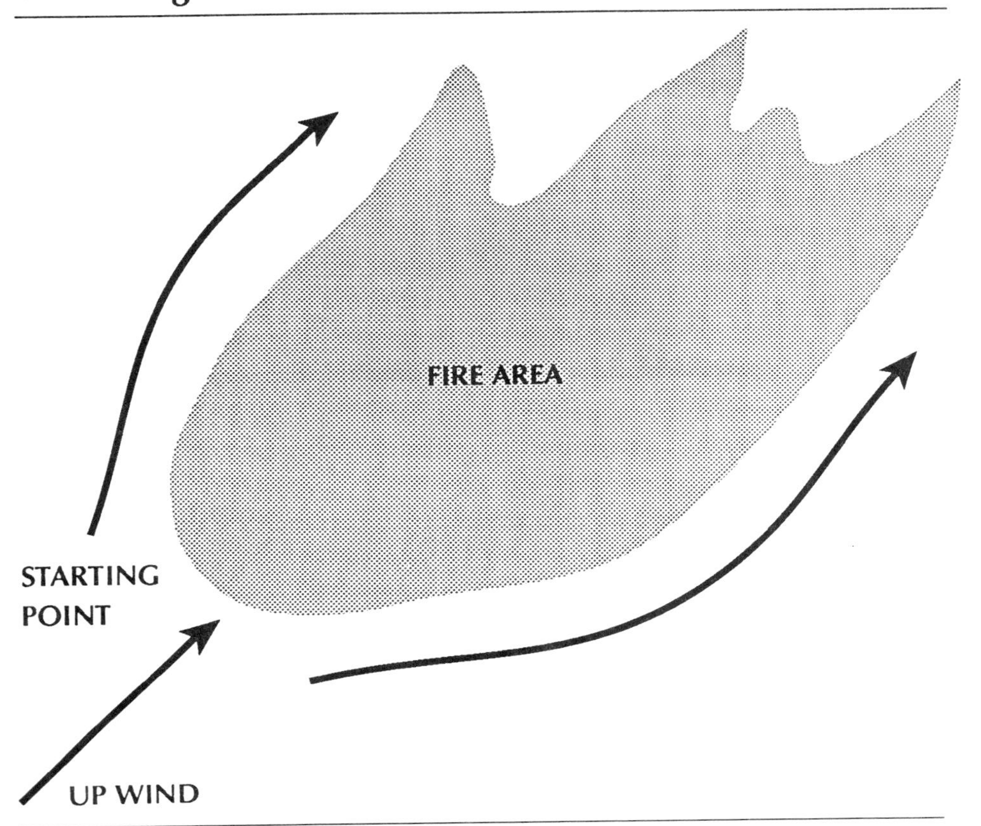

Figure 10.10. *The head of a fire can be approached safely from one or both flanks, and the fire cut off.*

should be undertaken only when life or property is endangered and an escape route is available to firefighters. Firefighters attempting this defensive operation should know that even with a pumper or brush vehicle and hose stretched and charged, they may not always be able to stop the oncoming rush of flames.

Convection currents of hot air, accelerated by a slope, or radiated heat waves, from flaming brush, may force firefighters to abandon the hose lines. Firefighters should also know that they cannot outrun the flames racing toward them, and they cannot run through the oncoming flames to the blackened area of safety. The best protective clothing will not protect them from the heat and flame. Unless an escape route is available, they will be trapped and burned to death. Bodies of firefighters attempting frontal attacks on brush fires have been found in roadside ditches, beneath pumpers on the ground where they have scurried for protection, and in nearby woods where they had become disoriented and lost. They were killed by the heat, the flames, or the lack of oxygen.

Firefighters with a pumper or brush vehicle sometimes become overconfident. These large vehicles, capable of high speed and large water delivery, can provide a false security. During the height of a blaze, the flames of burning brush can deplete the oxygen in the air, causing the pumper to stall, and a ride out of the fire area is then no longer available.

The water tank capacity of 500 or 1,000 gallons and the hose stream reach of 30 to 40 feet are ineffective against radiated heat or hot convection wind currents moving ahead of flaming brush. If a pumper becomes disabled in the woods—upwind or uphill of a serious brush fire—it should not be abandoned, and firefighters should not try to outrun the flames. Instead, as a last resort, the trapped firefighters should radio for assistance, stating their location; assemble inside the enclosed cab; cover themselves beneath canvas covers; close all windows, don SCBA, and wait for the flames to pass over the disabled apparatus. This procedure is a last resort and should be avoided. An orderly withdrawal is more effective than a disorderly entrapment.

Injuries

There are three kinds of injuries typically occurring to firefighters during brush firefighting: eye injuries, falls, and heat exhaustion.

EYE INJURIES

When fighting fires in brush and moving rapidly through heavy undergrowth or thickets, branch ends, vines, and leaves frequently whip across a firefighter's face. This shrubbery may often bear pointed needles, razor-sharp edges, or abrasive surfaces. Eye punctures, tears, and cuts have blinded firefighters battling brush fires. When a group of firefighters are heading through brush to cut off a fire, the lead firefighter is exposed to slice wounds by vines and needles as he opens up the path in the undergrowth. Firefighters walking behind the lead firefighter are, however, subject to eye injuries from the branch ends snapping back into their faces. To reduce the danger of eye injuries when fighting fire in brush, the lead firefighter cutting a path through the growth should raise a tool or arm in front of his face as he moves forward; eye shields of helmets should be lowered over the face; firefighters behind the lead firefighter should space themselves several feet apart.

FALLS

Firefighters in woodlands are not walking on a flat surface. The ground cover of fallen leaves conceals many hazards that can cause tripping: tree roots, stumps and logs, rocks, crevices, and holes. Firefighters rushing through brush to cut off a spreading fire, carrying heavy tools and equipment in both hands, trying to avoid sharp branch ends whipping and scraping across their faces, are in danger of tripping over a hidden tree stump or falling into a hole in the ground filled with leaves. If the falling firefighter is carrying a 40-pound water tank on his back, a serious injury may result.

A more serious hazard to a firefighter combating a fire in an open field or woodland is the presence of old, abandoned water wells, mine shafts, and caves. These deadly traps, sometimes covered over with rotted wood or leaves, could entomb an unsuspecting firefighter.

Figure 10.11. *Heat exhaustion is a common injury suffered by firefighters combating wildfires.*

HEAT EXHAUSTION

A firefighter's turnout clothing weights approximately 25 pounds, self-contained breathing apparatus weighs more than 30 pounds, a length of dry 1¾-inch hose weighs approximately 30 pounds, and a portable, five-gallon water tank weighs approximately 40 pounds. A firefighter operating at a brush fire, carrying 60 or 70 pounds of equipment on a hot summer day, walking in and out of a smoky wooded area to fight fires and to refill the water tank several times, can become exhausted (Figure 10.11). Heat exhaustion is a common injury suffered by firefighters combating wildfires. Training firefighters to pace themselves and providing rest periods, a refreshing water supply, and allowing opportunity for removal of turnout clothing, when safety permits, are necessary measures required at brush fires of long duration on hot days.

Lessons to Be Learned

1. Priorities for fighting wildfires are different from the firefighting priorities of structural fires (Figure 10.12). Structural firefighting priorities are: Protect life first, contain the fire second, and conserve property third. Wildfire firefighting priorities are: Protect life first, conserve property second, and contain fire third. When a wildfire does not endanger life or property, sometimes the best strategy is to let it burn.

1 Protection of Life
- **(a) Civilian**
- **(b) Firefighter**

2 Property Protection

3 Fire Containment

Figure 10.12. *Priorities of fighting wildfires.*

2. Firefighters should not operate uphill from a brush fire. A fire spreading uphill burns with the same speed and intensity as a fire fanned by a strong wind. The rate of flame spread in a brush fire on a mountainside increases with the steepness of the slope.

Figure 10.13. *A downwind current caused by a low-flying helicopter can cause a flare-up of blazing brush.*

3. The area upwind from a brush fire is the most dangerous area for a firefighter.
4. A flare-up—the sudden increase of flame from a small ground fire to a large area of blazing brush—may be caused by a gust of wind, a shift in wind direction, a downward current caused by a low-flying helicopter (Figure 10.13), or a change in the slope of the ground.
5. Most firefighter deaths and injuries occur in relatively light fuels such as grass, brush, and bushes (Figure 10.14).
6. Most firefighter deaths and injuries occur during relatively small brush fires or in isolated sectors of larger fires.

Figure 10.14. *Firefighters operating master streams may not always be able to stop an oncoming rush of flames from a wildfire.*

11

Aerial Ladder Climbing Dangers

THE FIREFIGHTER WHEELS the ladder apparatus around the corner and heads it into the fire block. It is still dark. The street lights are barely visible through the smoke banked down between the three-story, brick row houses that line the thoroughfare.

From a distance, the two firefighters in the apparatus can see flames blowing out of a pair of double windows on the first floor of the fire building. As they get closer, they notice smoke pushing out around a second-floor window frame and from an open window on the third floor. Smoke is also drifting downward from the decorative cornice at the roof level.

An uncharged hose line, connected to a pumper, leads up the stairs and into the front door. The chief stands in the middle of the street, waving on the ladder truck and pointing to the upper floor of the fire building.

The chauffeur brings the apparatus to a halt and looks up at the burning building. A young woman's head and right arm hang out a third-floor window. She waves frantically. Her head droops below the windowsill to escape the smoke that is pouring out above her. The wind momentarily changes direction, and she is no longer visible. Smoke now obscures the entire front of the building.

As the chauffeur slowly moves the rig into position for the rescue, the chief runs to the cab and shouts, "Did you see that woman at the window?" The chauffeur nods his head in an exaggerated manner as he concentrates on lining up the aerial ladder turntable with the window.

Preliminary controls and stabilizing systems are set for safety. Up at the turntable, the firefighter raises the aerial ladder from its bed, rotating it in toward the building. The aerial sections slide up into the smoke toward the window where the woman was last seen; it seems as though they will take forever to arrive there. The wind changes direction again—the woman is still at the window.

The second firefighter, wearing a ladder belt, is poised at the turntable for the climb. The aerial is lowered in toward the building just below the window, and the lever that locks the rung is applied. The firefighter springs up the ladder.

Down below him, firefighters are running into the fire building. Some are dragging a second hose line. The first hose line suddenly jumps like a snake as it is pressurized with water. Sirens of incoming units scream in the distance. The sound of steel striking steel and wood echo through the street as firefighters force entry into apartments.

Smoke blows into the face of the firefighter who's now halfway up the aerial. He looks up. The woman has already started climbing out of the window on to the ladder. "Stay there! Don't climb out!" The firefighter shouts. He climbs faster, looking into the smoke-blackened second-floor windows as he goes. He sees an orange glow through a cracked window pane. The fire has spread to the second floor.

As he reaches the woman at the tip of the swaying aerial ladder, fire suddenly explodes out of the window directly below them. Flames rise up and engulf the tip of the aerial. The firefighter quickly pushes the woman back into the window and dives inside after her to avoid the flames.

They scramble to their feet, and the firefighter grabs the woman by her shoulders. She is shaking violently and her eyes are glazed.

"Where's the door to the stairs?" he asks. She points. They crouch down below the smoke and make their way to the rear of the apartment. Flame crackles on the other side of the door. He opens it slowly. Smoke and heat pour into the apartment, and he quickly slams the door shut.

"Is there a rear fire escape?" he asks. The woman leads him across the room. He flings the curtains out of the way, opens the window, and peers down into the yard. Flames leap out of a lower window, engulfing the fire escape. There is no fire escape ladder to the roof.

The firefighter closes the window, and they move back to the front room. Through the window they see flames rising through the rungs of the aerial. Terror overcomes the woman. She turns to the firefighter with horror in her eyes. "We are going to die! We're trapped!"

Just then, a blast of water strikes the outside wall of the building and splashes through the open window, on to their faces. The crew members below have put a deck pipe into operation in an attempt to drive the flames back into the second floor. The firefighter feels the vibration of the stream as it hits the ceiling below. He looks out the window to the street; the chief is signaling frantically for them to climb down. He notices that the deck pipe is being supplied by a pumper's booster tank, not a hydrant—they have about two minutes to get out of the window and down the aerial ladder before the booster tank runs out of water and the flames again start to blow out the window.

Wasting no motion, the firefighter moves to the tip of the swaying aerial. The woman follows him to the window. He instructs her to sit on the sill, facing outward with her legs hanging over the ladder, then to roll over on to her stomach. He directs her right foot to the second rung of the ladder. Once she feels the stability of the ladder, now pressed against the building, she transfers her weight from the windowsill to the ladder rung. The firefighter, leaning over the woman, places her left hand on the aerial railing—just as the deck pipe runs out of water. Smoke and heat rush out of the window below and engulf them both.

"Help, I'm blinded! I can't breathe! I am going back inside!" the woman cries.

"No!" the firefighter commands. "Just step down! Three rungs and you'll be out of the smoke! Don't stop now!"

They move slowly down the aerial. The firefighter grips the side rails and maintains body contact with the woman, offering instructions and encouragement through the smoke. "That's it. Now take your right foot and place it on the next lower rung. Great!" A blast of flame surges up out of the window below, just above the woman's head. "Hurry—one more step!" the firefighter urges. "That's it."

They clear the smoke just as another ball of fire rises up the side of the building to engulf the tip of the aerial. The woman stops. "I'm afraid," she says. "We're too high up!" The firefighter is firm, yet calm. "Don't look down. Look at the rungs directly in front of your face. I will guide your feet."

Out of danger now, they climb slowly down the rest of the way, rung by rung. The firefighter guides the woman to the edge of the turntable deck, then carefully down the steps to the running board, and finally to the street.

The firefighter glances back up at the building. Flames are now blowing out of the third-floor window.

Most articles about aerial ladders are written from the point of view of the chauffeur or operator, the firefighter who positions, secures, and raises the aerial ladder at the fire. This chapter is different: It is written from the perspective of the firefighter who has to climb the aerial but has no hand in its positioning, securing, or raising.

At many fires, the aerial has already been raised. It may have been positioned at the window or roof perfectly; it may also have been raised at a difficult angle due to parked cars or overhead wires. Regardless of how it is positioned, when ordered to climb an aerial to accomplish a firefighting tactic, you must climb it.

Firefighters who climb ladders must realize that falls are a leading cause of fireground death and injury. True, most injuries caused by falls happen when firefighters are operating at ground level; however, the Bureau of Labor Statistics reports that 40 percent of on-the-job firefighter falls occur from elevations—these are the deadly, disabling falls.

Falls from portable ladders happen most often when the ladder moves or slips while the firefighter is on it. Falls from aerial ladders are different. They are often prompted by some event that causes the firefighter to lose his grip. Explosions, collapse of burning cornices, falling shingles, people jumping from windows, and dropped tools can all cause a firefighter to lose his grip and fall from an aerial ladder.

Ladder Belts

The firefighter who climbs an aerial ladder should wear a ladder belt around his waist. He may or may not use it; however, it should always be available if needed. It should be used on an aerial ladder when performing operations that require the use of both hands, if the ladder angle permits. (The ladder belt can be difficult to use when the aerial is positioned at a low angle—for that reason alone, ladder belts should be redesigned specifically for aerial use in order to provide greater flexibility of motion at any angle.)

Some firefighting duties that require the use of both hands are: stretching surplus hose to an upper floor, operating a ladder pipe, using a pike pole to open up a smoldering cornice, and assisting a person out of a window. A leg lock should not be used on an aerial as a substitute for a ladder belt. In fact, on some aerial ladder rungs, use of a leg lock is not even possible.

Climbing Technique

There is no one correct technique for climbing an aerial ladder; however, a continuous grip on some part of the aerial must be maintained. Feet must be positioned carefully on the rungs. The climb should be accomplished slowly, with leg muscles providing the thrust for upward progress. Rungs should not be skipped.

When the aerial is raised at a steep angle, gripping the rungs is often more comfortable. Gripping the rails is usually more comfortable for low-angle aerial placement (Figure 11.1). The hand holding a tool that might be carried up the ladder should be slid under or outside the railing, thereby holding the firefighter in toward the ladder.

There is always a danger of falling when descending an aerial ladder. Firefighters should always face the rungs—face the building—when climbing down, even when the aerial is positioned at a low angle. That way, if a firefighter misses a rung, loses his balance, and trips, he will fall into the ladder and have a better chance of regaining handhold and footing. Facing away from the aerial in descent will, if a rung is missed, propel the firefighter outward, and his chances of regaining any grip on the aerial are small.

Figure 11.1. *Climb an aerial ladder and grip either the rungs or the side rails, whichever is more comfortable. Maintain a continuous grip on the ladder.*

Figure 11.2. *The air tank of a mask can cause a firefighter climbing a ladder to fall backward; the weight of the equipment changes a firefighter's center of gravity.*

Firefighters operating with self-contained breathing apparatus and carrying tools are more likely to fall from aerial ladders. The 30-pound cylinder strapped to a firefighter's back can cause an error in balance; this extra weight changes the vertical equilibrium of the body (Figure 11.2). A heavy tool, such as a power saw, may have the same effect. In addition, the effort used to hold on to it may interfere with the firefighter's concentration.

Tools should be attached to slings or straps whenever possible, allowing them to be carried over the shoulder. Both hands, then, will be free to grip the aerial. Many aerial ladders are designed to hold an axe or pike pole in a bracket attached to the tip of the aerial, leaving the firefighter's hands free to use them for the upward climb.

Climbing While Raising

There's plenty of photographic testimony—published and unpublished—to this dangerous practice. We've seen pictures of firefighters grimacing in pain because their feet and ankles were crushed between ladder rungs that moved while they were on the aerial ladder.

Firefighters should never be standing on an aerial ladder while it is being raised, rotated, or extended. The aerial must be in position before any firefighter climbs it, which includes making sure that the ladder locks are set, too.

If a firefighter climbs the aerial ladder while it is being raised or moved to reach a known victim more quickly and his foot is caught between moving ladder rungs, the rescue attempt will have to be aborted and will probably never be accomplished. The rescue scene will show a firefighter caught on the ladder, screaming in pain; the aerial hanging in midair; and the victim trapped at the window. Unless another aerial ladder at the scene can be quickly moved into position, the trapped victim will be forced to jump.

From Aerial to Window

A hazardous moment during an aerial ladder climb occurs when the firefighter moves from the tip of the raised aerial into a window that's high off the ground.

He should first place any tools he is carrying inside the window opening. With both hands free, he should then grasp the sides of the window or the sill and test its stability. If it does not move, the firefighter should maintain this grip and pull himself up through the window (Figure 11.3). He should not grab any part of a burned or charred window frame, which might crumble in his hand.

Figure 11.3. *A firefighter can fall when climbing through a window of a building from or to the tip of a ladder.*

When climbing from the building back to the aerial after the operation has ended, the tools should be placed near the window where they can be easily reached. While holding on to some secure part of the window or building, the firefighter should climb on to the ladder, maintaining a grip until he feels his weight being supported by the aerial. Only when balanced and fully supported by the ladder, should he pick up the tools, secure them in a sling or strap, if available, and climb down the ladder—maintaining a firm grip at all times.

Dangers of Shaftway Window Openings

In commercial buildings, some windows facing on the street front open into elevator hoistways. There is no floor inside these windows, only a 50- or 60-foot drop down an elevator shaftway. Firefighters have climbed from aerial ladders into shaftway windows and fallen to their deaths. Four windows may be placed in a row across the width of a storage building. Three may give access to the floor, but one may open into an elevator shaft.

In some communities, the law requires a warning sign on all shaftway windows; however, these signs, if wooden, burn away when flames explode out the windows at a multiple-alarm fire, may be obscured by clouds of smoke, or may simply be unreadable at night (Figure 11.4).

Always remain aware of this very common hazard in commercial occupancies. Tools can come in handy at such operations. Placing tools inside a window before moving from the aerial to the building, in addition to freeing both hands, tells you if there is a floor deck or not; if you don't hear or feel the tools strike the floor, you may be confronting a very long drop down an elevator hoistway.

Aerial to Flat Roof

When an aerial ladder is placed several feet higher than a roof parapet, the firefighter charged with rescue, vertical ventilation, or some other tactic should stop at the parapet wall and drop his tools on to the roof. He then continues to climb several more rungs until he can easily step over the parapet wall. He should always grip the ladder when moving to the roof.

If the tip of the aerial ladder is placed below the parapet or roof, the firefighter will have to grip the parapet or roof cornice during the move from aerial to building. This technique is more dangerous.

Figure 11.4. *To determine if the floor inside a window has not burned away or to ensure that you are not climbing through a window into an open shaftway, drop your tool on the floor before entering the window, and listen for the tool to strike the floor.*

Masonry parapet walls have often deteriorated and been weakened by the elements. Wood cornices are often rotted and decayed. Firefighters must test the stability of the parapet before climbing to the roof.

Assisting a Person from a Window Down an Aerial

Assisting a trapped person from a window of a burning building to the tip of an aerial ladder is one of the most dangerous and difficult ladder operations a firefighter may ever be required to perform during

Figure 11.5. *Taking a victim down an aerial ladder may have to be done at night, during a rain- or snowstorm, and at great heights.*

his career. It may have to be done in the dark of night, during a rain- or snowstorm, and at great heights (Figure 11.5). It may have to be accomplished quickly, before a smoke-filled room explodes into flames. During this action, the firefighter may be required to maintain his balance while holding the victim, who quite possibly has never climbed a ladder, and most definitely is scared and excited.

Moreover, sometimes this dangerous feat need not be performed at all. As always, the firefighter must size up the situation before acting. The decision to take a person down an aerial may not be automatic.

The firefighter must first determine if the danger from the smoke and fire are greater than the danger of the climb down the aerial. If the victim is hanging out the window and grasping desperately at the sill, or is standing on a ledge of the building next to a flaming window, there is only one decision to be made. If, however, the person is trapped several floors above a fire, or there is only a light smoke condition in the room, the victim need not be exposed to the danger of the aerial ladder descent.

The firefighter may be able to climb into the window and calm the person, shut the door to the room, stopping the entrance of smoke, communicate the situation to his officer by portable radio, and stay with the victim at the window until the fire below has been extinguished.

There are priorities that dictate how a victim should be removed from a burning building. These priorities are based on safety—victims must be removed from burning buildings in the safest possible manner. The escape-route priorities, in order of preference, for victim removal during a fire are: smoke-proof tower, interior enclosed stairway, a safe fire escape, an aerial platform, and an aerial ladder. A victim should be taken down an aerial ladder only if none of the preferred avenues of escape is available for use.

When a firefighter makes the decision to remove a person from a burning building by means of an aerial ladder, the following safety procedures should be considered:

- The firefighter should be out on the aerial.
- The person should be instructed as to how to climb out of the window.
- While the victim is holding on to the windowsill, his feet should be guided and placed on the rungs of the aerial ladder (Figure 11.6).

Figure 11.6. A person's hands and feet should be guided and placed on the rungs if necessary. Several firefighters may be needed to remove a person safely from a burning building down an aerial ladder.

- The victim should then be encouraged to transfer his body weight to the aerial, then to move one hand from the windowsill to the rail, and then the other.
- If the fire is extending out the window during rescue, very little instruction will have to be given; the person will move automatically from the windowsill to the ladder. If, however, the victim freezes on the ladder, he should be told that he must not look down, but instead must look up or concentrate on the rungs in front, using a foot to feel for each rung below him. If necessary, the firefighter could guide and even place the victim's feet on each rung at the beginning of the climb down the ladder.

Firefighters who have successfully guided individuals down aerial ladders report that they continually talk to the victims during the descent. They give encouragement and congratulate them on their progress. Once the person is moving down each rung without much hesitation, the firefighter can maintain body contact with the victim by enveloping the person with his arms while holding on to the ladder rail; this position can give a sense of security on the aerial. Before each aerial section is approached, the firefighter should warn the person that the rung spacing will change. The person's foot may have to be guided to these rungs.

- Victim removal down an aerial ladder is not completed until the person is off the apparatus, at street level, and in the care of medical personnel. People have been guided down a 75- or 100-foot aerial ladder, only to be left on the turntable to faint and fall eight or ten feet to the ground. Individuals taken down an aerial ladder should be assisted off the apparatus; they must never be allowed to wander off in a state of shock.

Turntable Hazards

There are certain hazards and dangers that the firefighter must consider when he nears the turntable platform after climbing down a fully extended aerial ladder. He will be exhausted after performing search and rescue, ventilation, or extinguishment tactics and may tend to become careless. The urge to skip bottom rungs of the aerial is great.

On some metal aerial ladders, the distance between the last two rungs varies. Furthermore, the shape of the last rung is different: there is a large metal drum with a greased cable which must be stepped over before you reach the turntable platform. Even if you reach the platform safely, climbing backward down a fully extended aerial ladder, with your head down while concentrating on and facing the rungs, can cause problems in balance. One step backward over the circular edge of the turntable could result in serious injury. Furthermore, just the seemingly simple act of climbing from turntable to street can be dangerous for those not assigned to a ladder unit—the steps and metal foothold are sometimes not recognizable to them. Firefighters have been injured by stepping on a raised compartment door. Don't take the short climb from turntable to street for granted—it may not be as simple as it seems.

Lessons to Be Learned

Climbing an aerial ladder is a dangerous task that requires skill, agility, knowledge, and discipline (Figure 11.7). One of the major responsibilities of a trained, qualified chauffeur assigned to position an aerial ladder is to ensure that all safety devices are set in place before anyone climbs it. When all is in position, the climbing firefighter must do so slowly and carefully.

Safety must always take priority over speed. Dependability, not speed, is the principle on which effective, safe firefighting is based.

Figure 11.7. *Climbing an aerial ladder is a dangerous task. It requires skill, agility, knowledge, and training.*

12

Forcible Entry Dangers

AROUND MIDNIGHT, an engine and a ladder truck arrive at a new shopping center. Smoke fills the parking area in front of a row of ten new stores. Each is 20 feet wide, 40 feet deep, and two stories high, with retail areas on the first floor and offices on the second.

Protecting each store's large display window is a steel-mesh security gate rolled down from its horizontal metal housing and held in place by two padlocks—one on each side.

Large quantities of smoke are pushing out of the top of the display windows of the five center stores.

"Ladder 6 to communication center," the truck company officer calls into his apparatus radio. "Transmit a second alarm for a working fire in the Kings Point shopping mall."

The pumper lays a supply line from a hydrant in the parking area, and the firefighters stretch a 1¾-inch pre-connect attack line to the front of the stores. The ladder apparatus is placed away from the front of the stores, upwind of the fire. The ladder company members spread out, peering into each smoke-filled store and trying to locate the main body of fire.

On one of the stores, a black-gray film, streaked with condensation, coats the inside of the two large, plate-glass display windows. The windows are pulsating in and out, and smoke puffs out around the glass door set back between the two windows.

"Hey, Cap, I found the fire! Here it is," shouts a firefighter standing in front of that store.

All the firefighters regroup there, and the captain orders them into action. "You," pointing to one, "get a ground ladder; check out

Figure 12.1. Firefighters performing forcible entry tactics are endangered by backdraft explosions.

the floor above. You," pointing to a firefighter who has a portable radio, "get to the rear of this store. See if you can vent. When the engine gets the line in position, I'll give you the word." Turning to the remaining firefighters, the officer orders, "Force that security gate open" (Figure 12.1).

The firefighter carrying the axe and halligan tool sizes up the gate. The padlocks are cheap and can be forced easily. He hands the other firefighter the eight-pound flat-head axe, places the point of the halligan tool in the bow of the padlock on the left, and slides his gloved hands down to the forked end of the tool.

Pulling at the outstretched padlock, he orders the other firefighter to swing the axe. The back of it strikes the halligan tool behind the point placed through the padlock bow. *Bang!* The padlock breaks away from the staple and lock pin which it held secure and falls to the sidewalk.

Quickly moving to the other side of the security gate, the team repeats the procedure. *Bang!* They knock the other padlock off.

The officer's radio blares. "Ladder 6 rear to ladder 6."

"Ladder 6. Go ahead."

"Captain, there's no way to vent the rear. It's a solid brick wall."

"Ladder 6, ten-four."

Walking over to the engine officer, the truck captain says, "When we get this gate up, we're going to take out the display window. Let the fire blow; then you guys can move in. There's no way to vent at the rear or above, so watch it."

Inside the recessed entrance, the glass door suddenly cracks because of the heat in the store and crashes to the floor behind the security gate. Smoke momentarily billows out and envelops the firefighters, but it retreats almost immediately, sucked back through the broken door.

Boom! The captain is lifted off his feet, blown backward a yard or so, and slammed on to his back, skidding and sprawling in the road.

Slowly, very slowly, he rises up on one knee, shaking his head. Staring at the front of the store, he can't believe his eyes. Flames shoot out a gaping, black hole where the display windows stood seconds ago. Several firefighters are getting up off the ground and running over to the hose line, where a firefighter without a helmet is directing the nozzle into the flaming store.

Small pieces of broken glass are scattered over a 50-foot radius in the front of the store. The 20-foot metal ground ladder has been blown over and lies twisted, bent 90 degrees out of line at its center, one end stabbing crazily straight up into the night air. The entire security gate, together with its frame and roll bar housing, has been blown off the front of the store in one piece, and now it's half in the road.

"I'm alive," the captain thinks. "What an explosion!"

"Captain, captain!" A firefighter runs over to him. "What happened? I was blown off the ladder, but I'm all right. God, this is something!"

"It must have been a backdraft," the officer manages to reply, still dazed. "Where's the forcible entry team?"

Just then, they hear a cry. "Help! Help! Get this off me!"

The captain and the firefighter run over to the security gate and lift it several feet. A firefighter crawls out from beneath, his face covered with blood. He tries to stand up.

Figure 12.2. TOP: *Backdraft or smoke explosions can occur in adjoining occupancies. The store to the left of the fire will explode.*

Figure 12.3. MIDDLE: *Firefighters ready to remove the stucco-covered plywood wall will be injured by a backdraft smoke explosion that is about to occur.*

Figure 12.4. BOTTOM: *The stucco-covered plywood wall exploded outward, burying firefighters. Other firefighters rush to their rescue.*

"Joe's still under there," he whispers.

"Quick, get that portable ladder and wedge it under this gate while I hold it," the captain orders the other firefighters. They turn the bent ladder on its side and place it under the gate. The captain crawls under the metal screen, drags out the unconscious firefighter, and rolls him over. Blood pours from a wound in the man's chest.

"He fell on the halligan tool; the point punctured his chest. Get an ambulance. Hurry!"

A backdraft is a combustion explosion or smoke explosion, which is defined as the result of the chemical reaction of heat, oxygen, and fuel—the same ingredients that create a fire (Figures 12.2, 12.3, and 12.4). (The other broad classification is a physical explosion—the rupture of a cylinder or container because of excessive pressure.)

An explosion differs from a fire, though, because the speed of the reaction and the rapid expansion of gases cause shock waves. A fire doesn't cause them.

Not all explosions occurring at fires are backdrafts. Most are caused either by natural gas leaking from broken or melted gas piping during the fire or by residual vapors of the flammable liquids an arsonist has used to start the fire.

Explosion Dangers

1. **Gas Meters**
2. **Vapors from Arsonist's Flammable Liquids**
3. **Propane Cylinders**
4. **Household Pressurized Containers**
5. **Window Air Conditioners**
6. **Kerosene Containers Used for Space Heaters**
7. **Imploding Television Tubes**
8. **Smoke Explosions Due to Double-Paned Replacement Windows**
9. **Backdrafts**
10. **Gas Piping in Turn-of-the-Century Houses Which Used to Serve Lighting Fixtures**

Figure 12.5. *The above list gives some causes of explosions in burning buildings.*

Serious blasts from the rupture of cylinders containing liquefied petroleum gas are also becoming common, while most minor puffs of smoke at fires are caused by household aerosol containers.

Before a fire officer declares that an explosion was a backdraft, an investigation should be conducted in order to eliminate these other types of explosion (Figure 12.5).

Firefighters often ask what the difference is between backdraft and flashover. There are several.

A flashover is not an explosion. Both backdraft and flashover are rapid combustion reactions, but the backdraft chemical reaction happens faster—so swiftly that the suddenly heated atmosphere expands and creates excessive pressure in the fire area. This pressure is released or dissipated as shock waves. Flashover, the sudden ignition of combustible fire gases, doesn't produce shock waves.

Another difference is that a flashover occurs only during the first stage of fire development, the growth stage (Figure 12.6). A backdraft occurs during the growth or decay stage of a fire.

Finally, the triggering events are different. For a backdraft, the triggering event is the addition of oxygen, the fresh air introduced to the confined area after forcible entry (Figure 12.7). The triggering event for a flashover is the addition of heat, caused by radiation feedback to the fire area (Figure 12.8). This heat is generated by the fire, accumulated at the upper levels of the fire room, absorbed into the ceiling and walls, and re-radiated into the room. There it ignites the fire's combustible gases in the presence of sufficient oxygen.

Safeguards Against Explosion

Scientists state that there are five ways to safeguard against an explosion or its effects: contain the explosion; vent its effects; quench the explosion with an extinguishing agent; isolate it from people; or "dump," or remove, the explosive material.

Firefighters can't remove the explosive material, but they can use the other four procedures to protect themselves against a backdraft during forcible entry.

Containment

When sizing up a door for forcible entry, a firefighter checks to see if it is locked, analyzes the type of lock, determines if the door opens inward or outward, and feels the door to estimate the amount of heat and fire behind it. If there's intense heat behind the door and if the

Time-Temperature Course of a Fire

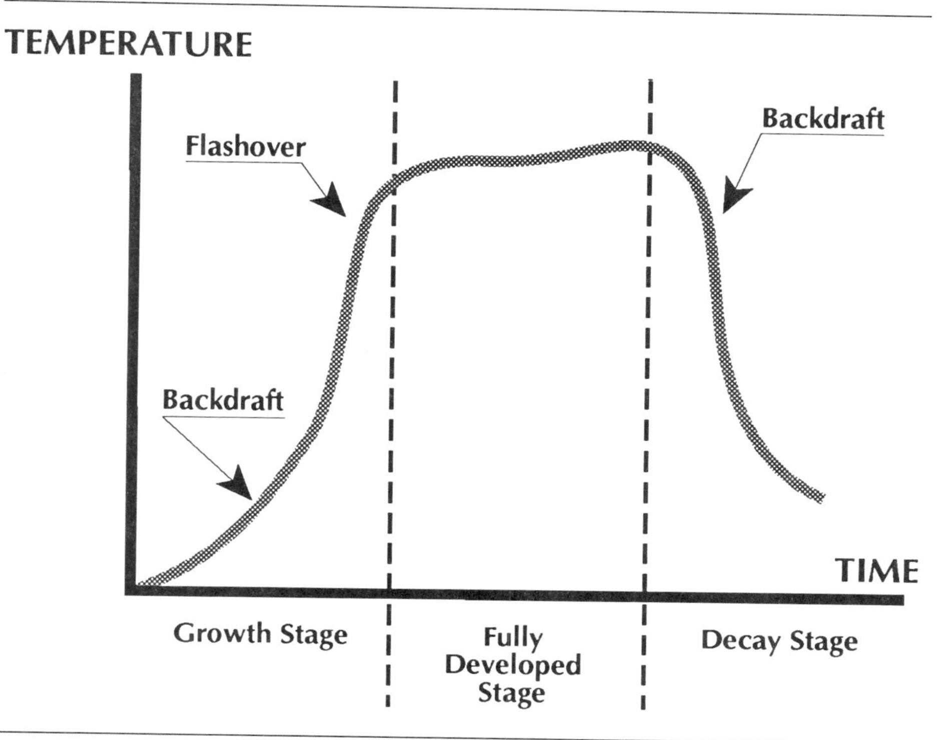

Figure 12.6. *A flashover occurs in the growth stage of a fire. Smoke explosions can occur in the growth or decay stage of a fire.*

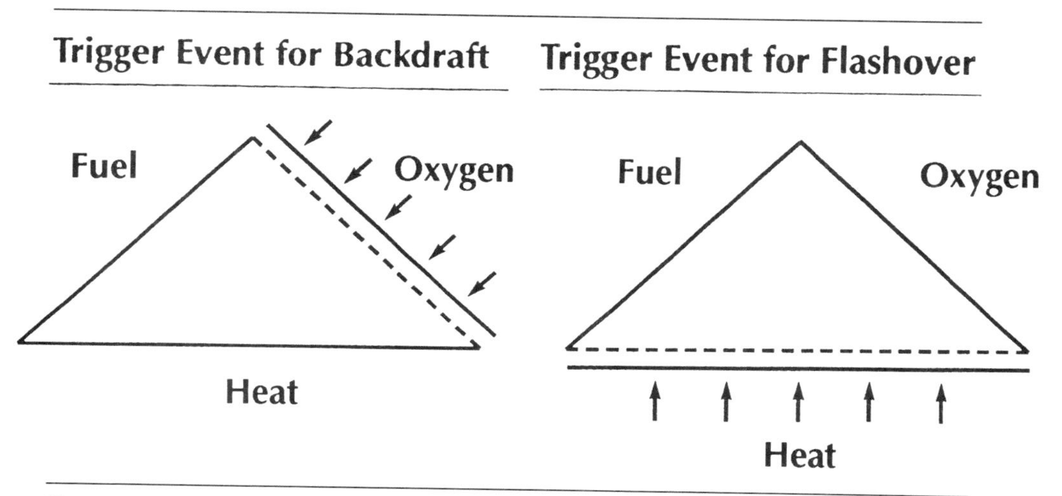

Figure 12.7. LEFT: *Oxygen in the air triggers a backdraft explosion.*
Figure 12.8. *The build-up of heat in a room triggers a flashover.*

Figure 12.9. The inward swing of a door must be controlled in order to prevent inrushing air from triggering a backdraft explosion: containment.

door swings inward, the door must be held closed or its inward swing controlled in some manner during forcible entry in order to prevent air from rushing into the fire area (Figure 12.9). This position could "contain" the explosive atmosphere.

To accomplish this task, a firefighter or the officer should crouch below the members of the forcible-entry team, who are standing, and hold the doorknob closed with a gloved hand or a short piece of rope.

Venting

Backdrafts often occur in shopping centers and rows of stores which are closed overnight or over weekends, where fires can remain undetected for a long time. These structures are usually one-story buildings which can be vented at the roof, and vertical ventilation above a fire, before the front door is forced open, happens to be one of

Figure 12.10. Vertical ventilation is one of the most effective methods of preventing or reducing the explosive effects of a backdraft: venting.

the most effective methods of preventing or reducing the explosive effects of backdraft (Figure 12.10).

Where glass skylights are present on a flat roof and are located directly over the fire, they should be immediately removed to vent the fire below. This action will release any superheated gases upward and outward, offsetting the chemical reaction of heat, fuel, and oxygen inside the store. Even if backdraft occurs, some of the shock waves will be diverted out the skylight.

Quenching

Unfortunately, skylights have almost disappeared from the roofs of one-story retail stores during the past 30 years. Store owners have sealed up skylights which a burglar might use to enter, so quick venting to avoid backdraft is becoming rare. Roofs must now be cut with power saws and the ceiling below pushed down with pike poles in order to vent a store fire vertically. Firefighters usually force the front doors open before this slower form of roof venting can be done.

In any case, roof venting cannot prevent backdraft in a two-story structure when the fire is on the first floor.

Therefore, another protective method must sometimes be used. A charged hose line must be ready at the front door when the door is forced. A quick stream of water on a superheated fire area, as soon as the door is open, could cool a potentially explosive atmosphere (Figure 12.11). This method isn't as effective as roof venting, but it's a practical alternative today.

Isolation

In some instances, the fire inside a store is at the rear, behind a partition wall, and the hose stream couldn't hit it from the front door. If there are no top or rear vent openings, the firefighters are, in effect, standing in the barrel of a loaded shotgun that's about to fire when fresh air enters the burning store. The full blast of the backdraft will be coming out the front entrance into the path of the advancing firefighters.

As a last resort, firefighters should break the glass display windows, then step back quickly as the fire or explosion vents out the front entrance. At that point, they can advance a hose line—already placed and charged ahead of time—into the now flaming store (Figure 12.12). This technique allows the explosion to occur in "isolation" and safeguards the firefighters from the blast. One disadvantage is that the fire may spread above and into the adjoining store; hose lines will therefore have to be positioned at these areas, too.

The tactics for dealing with backdraft are available for use once a proper size-up has recognized backdraft conditions early. That combination is the key to firefighter safety when it is a case of forcibly entering a tightly sealed building or room where fire may have been burning for a long time.

A less serious hazard to a firefighter assigned to forcible entry of a burning building is the risk of being struck by a tool when using an axe or halligan tool to force open a door. Firefighters holding halligan tools have had their jaws broken, fingers, hands, and wrists fractured, and have been stabbed in the neck and chest. Most of the injuries are caused by the firefighter swinging the axe in an uncontrolled or uncoordinated manner.

When two firefighters force a door using the axe and the halligan tool, the firefighter holding the halligan tool must assume leadership of the team and direct the forcible-entry operation. Size-up of the door, placement of the halligan tool, and prying of the door comprise

Figure 12.11. *A quick stream of water directed into a superheated fire area soon after forcible entry could cool a potentially explosive atmosphere:* quenching.

Figure 12.12. *Firefighters allow flame or explosive blast to vent out a store front while they are safely positioned to one side:* isolation.

the more complex of the two tasks. The more experienced of the two firefighters assigned to forcible entry should perform this duty of halligan tool placement and holding. This firefighter, after placing the halligan tool near the lock to be forced, must tell the firefighter swinging the axe when to strike the halligan tool and when to stop. He must also tell him when the halligan tool will be moved slightly to achieve a better placement in the doorjamb and when to stand back as the door itself is being pried open. There must be no surprises, no sudden moves by the firefighter holding the halligan tool, or he may cause his own injury. Every move by the firefighter holding the halligan tool must first be announced to the firefighter swinging the axe. "Strike, strike, strike." "Stop." "Stand back a second." "Okay, strike, strike, strike." "Stop." "Hold it." "That's it." There are some examples of the communications that must take place between the members of a forcible-entry team. Communication equals safety during forcible entry. After the order has been given to "strike," the halligan tool must be kept in one steady position. It cannot be moved even slightly, or the axe may miss or angle off the tool and hit the firefighter holding the halligan tool.

The firefighter with the axe does not actually swing the axe to strike a halligan tool forcing open a door. Instead, the back of a flat-head axe is moved forward forcefully in a punching action. One gloved hand holds the axe near the end of the wooden handle; the other hand should be placed just below the axe blade. The flat-head portion of the axe strikes the halligan tool with short, sharp blows, increasing progressively in force until the one holding the axe is ordered to stop. The power behind the axe movement comes only from one shoulder and the weight of the axe head. An eight-pound axe or heavier striking tool should be used for forcible entry in place of a six-pound axe; it is safer to increase the size of the striking tool than to increase the force of the swing. The firefighter using the axe to strike the halligan tool must remain constantly alert to orders given by the firefighter holding the halligan tool and must quickly react to them. When suddenly ordered to stop, the firefighter with the axe may instantly have to pull back the axe, in mid-swing, in order to avoid injuring the other firefighter.

The key to safe forcible entry operations using an axe and a halligan tool is coordination between the two firefighters, control of the operation by the firefighter holding the halligan tool, and practice. For safe operations during every tour of duty, the firefighter assigned forcible-entry duties must practice holding the halligan tool and swinging the axe that is striking the tool, as well as discussing various methods of forcible entry, various locks, and alternative tools available for use during forcible entry.

Venting Large Glass Windows

Another dangerous operation sometimes performed by forcible-entry firefighters is venting large plate-glass store windows. Firefighters have suffered severed arteries, cut nerves, sliced tendons, and lost the use of fingers and hands due to severe lacerations caused by broken glass. Firefighters have bled into unconsciousness after cutting open an artery or a vein. Wounds, cuts, bleeding, and bruises are the second leading group of firefighter injuries (sprains and strains are number 1).

There are several important differences between breaking a small residence window and breaking the large plate-glass window of a commercial structure to vent smoke from that building. Venting the large plate-glass window is more dangerous. For example, the glass from a residence window is usually only one-eighth inch thick; when it is broken, its pieces are small; its weight, when falling, is negligible. A firefighter caught below falling glass at a fire in a residence building can imitate a turtle, avoid looking up, and, if wearing proper protective clothing, can avoid serious injury. Not so with falling glass from the large plate-glass window of a store. Glass windows of commercial buildings are one-half inch or more in thickness; when broken, they fall in large sheets; furthermore, these broken pieces can weight 50 to 100 pounds. A 50-pound section of falling plate glass can decapitate a firefighter or knock one off a ladder. When ordered to vent a large glass display window of a store during a fire, a firefighter should use a pike pole six or eight feet long. The longer pike pole will provide more distance between you and the falling glass of the window's top portion. Stand to one side of the window to be vented. First, strike the glass window near the top section. If the glass were to be broken near the lower half, a large section of glass could slide down and cut the firefighter holding the pike pole. Firefighters should remove all broken glass pieces still stuck in the window frame after the window has been broken. A large section of plate glass left in the top of a store-window frame may suddenly fall like a guillotine.

When examining injury reports of cuts and wounds caused by broken glass, we see the same type of injuries occurring. Inexperienced firefighters suffer cuts and lacerations in performing the same firefighting acts as their predecessors: kneeling on broken glass without first pulling up their boots; climbing through windows where some glass pieces have not been removed from the window frame, breaking a window with a gloved hand holding a flashlight, breaking a large, plate-glass store window with an axe or halligan tool, placing a ground ladder below a window and then venting the glass window directly above. When firefighters are positioning ladders for venting, the ladder tip should be placed to one side of the window. After the window has been vented, the ladder should be repositioned directly below the window opening for entrance or exit.

Broken glass can cause injuries other than cuts. Broken glass pieces resting on top of one another in the street, on stoops, porch roofs, and fire escape landings must be considered a fireground hazard: broken glass can become as slippery as ice. Firefighters walking on a surface covered with broken glass can fall off porch roofs, off stoops leading up to an entrance door, and off fire escapes. Ground ladders resting on a sidewalk covered with glass can quickly slide away from a building while supporting the weight of a firefighter. A firefighter raising a ground ladder and setting the climbing angle must ensure that the butt of the ladder is not set on broken glass. In addition, a ladder must always be butted by another firefighter or be secured at the tip when someone is climbing. The front of a fire building is often littered with broken glass, which can first cause a firefighter to lose his balance and fall, after which the broken glass may cut him as he hits the ground.

Lessons to Be Learned

1. When firefighters talk about backdrafts, they invariably discuss warning signs. These signs include a reverse flow of smoke back into a doorway, smoke puffing out around a door frame or window, and the movement of a large plate-glass window in and out in a pulsating manner. Warning signs, however, are not the answer to fireground safety during a backdraft or an explosion; a firefighter should not believe he can avoid injury from a backdraft or an explosion by looking for warning signs. Explosions happen too fast—there is no time to react (Figures 12.13 and 12.14). The only protection a firefighter has against an explosion blast is his protective equipment: gloves, mask, helmet, turnout coat, pants, boots. After a blast, all exposed areas of the firefighter caught in the explosion may be severely burned (Figures 12.15 and 12.16). A firefighter who is properly equipped with all protective gear should be able to take the full blast of a backdraft. If he does

Figure 12.13. *Firefighters advance an attack hose line soon after forcible entry.*

Figure 12.14. *Backdraft explosion occurs.*

Figure 12.15. *Firefighters hastily retreat.*

Figure 12.16. *Injured firefighters are treated.*

not wear all the protective clothing, he will probably never again perform forcible entry or extinguish another fire.

2. Four safety precautions a forcible entry team can use to prevent or reduce the effects of a backdraft explosion are: *(a)* Control the door; do not let it open accidentally during forcible entry, because this motion will feed oxygen to the fire area. *(b)* Vent the roof quickly, if possible; this action could prevent or dissipate the blast of a backdraft. *(c)* Keep a charged hose line ready at the door. If venting cannot be accomplished and the fire is within reach of the hose stream, a quick discharge of water on the fire may inhibit a backdraft explosion. *(d)* When the fire is not within reach of a hose stream at the door and there is no possibility of venting, open up the front of the fire area, stand aside, let the superheated gases explode into flame, and then move into the fire area and extinguish the blaze.
3. When using an axe and a halligan tool to force entry, coordination, control, and practice are the methods to ensure safety and prevent injury. The firefighter holding the halligan tool should be the team leader and control the operation.
4. Wounds, cuts, bleeding, and bruises are the second leading cause of firefighter injury. These injuries most often occur during window venting operations to clear smoke from a confined fire area. Venting the large plate-glass windows of burning and smoke-filled commercial buildings is more dangerous than venting the windows of burning and smoke-filled residence buildings.
5. Remember, before forcing any door to a fire—try the door knob—it may be open.

13

Master Stream Operation Dangers

"COMMUNICATIONS TO BATTALION 1. Go ahead with your message."

"Battalion 1 to Communications at Box 214. We are going to use an outside attack on a fire in a three-story, 20- by 60-foot wood-frame vacant building at 2147 Webster Avenue. The structure is fully involved. We have fire on all three floors. We need an aerial platform to respond to this box."

"Communications to Battalion 1. You are getting Ladder 6, which is an aerial platform."

"Battalion 1 to Communications. Inform Ladder 6 while en route that the battalion wants them to set up in front of the fire building, and that they should approach the fire from Fordham Street, against traffic. Traffic has the street blocked."

"Communications to Battalion 1. Message received. The time is 1330 hours."

"Ladder 6 to Communications, We receive and acknowledge Battalion 1 message."

Ladder 6's chauffeur slowly moves the vehicle into the smoke-filled street. In the afternoon sunlight, smoke banks down into the street, reducing visibility. The chief runs to the cab of the incoming aerial platform and steps up to the running board on the officer's side.

"Captain, we have supply lines being stretched to the front of the fire building. I want your aerial stream in operation right in front of the building."

"Okay, Chief. Will do."

Flames are blowing from every opening on every floor of the vacant building's front. The asphalt siding on the building has caught fire. Flames travel vertically up the siding, while flaming droplets of melting asphalt rain down the building's face. The concrete sidewalk is a pool of burning, black liquid.

A wooden decorative cornice near the roof line is in flames. Suddenly it collapses at one end and swings down like a pendulum across the front of the structure. When the fallen cornice hangs straight down, the attached end breaks loose, and the flaming wooden structure crashes to the sidewalk.

As the turntable of the aerial platform is centered in front of the building, the chauffeur stops the vehicle, sets the brakes, engages the power take-off, and jumps out of the cab. At the aerial platform control panel on the side of the apparatus, the firefighter pulls a handle down to activate the corner jacks. The jack plates hit the pavement, and the truck is lifted into the air. Another firefighter places lock pins in each jack.

Now on the turntable, the chauffeur waits for the go-ahead signal from the captain and the firefighter who are climbing into the basket, fastening their safety belts. The chauffeur raises and rotates the boom from the trailer toward the burning building. He speaks into the two-way intercom between turntable and basket.

"Captain, take over the controls in the basket whenever you are ready."

"We are ready. I'm taking over the controls," replies the captain.

The men in the basket can feel the heat from the flames on their faces.

"We are going to start on the lower floors and work up," the captain says to his firefighters.

"Ten-four, Captain."

As the boom is extended toward the fire, they shield their faces from the heat waves, turning away from the burning building and pulling up the collars of their turnout coats.

"Start the water," the captain calls over his walkie-talkie.

"Here comes the water, Captain!" the firefighter on the turntable shouts into the intercom.

The firefighter in the aerial platform basket readies the deluge nozzle affixed to the front of the basket as the captain operates the controls. The stream of water spurts out of the 1⅝-inch nozzle tip. Slowly, water pours down into the street without pressure, then straightens out, suddenly and violently, smashing into the front wall of the burning building and cascading into the street. The firefighter gains control of the stream's direction and aims it into a ground-floor window, instantly driving the flames back into the structure.

"We are going to the second floor!" the captain shouts above the roar of the accelerating diesel engine, over the sound of the master stream hammering the front wall of the building.

The captain keeps both hands on the controls to ensure smooth movement of the rising basket. Because of the back pressure of the large-caliber stream, any erratic motion of the basket controls is magnified tenfold. Flames are visible through the top-floor windows. The captain looks over the basket railing to check for flames that could reignite below them. As he moves in toward the building, the nozzle of the aerial stream is inserted into the top-floor window near the sill and directed upward. The stream bounces off the ceiling; it extinguishes the flames.

The captain notices flames at the rear of the top floor; he moves the basket in closer for deeper penetration. Abruptly, without warning, the building collapses.

"Hold on!" the captain shouts.

The firefighter shouts back, "Captain, the nozzle is caught in the window; it's pulling the basket down!"

In an instant, the basket of the aerial platform and the extended boom are dragged down with the collapsing building. The nozzle unsnags from the window and whips violently back and upward. The basket springs back up and the firefighters grip the railing; but, as the basket reverses direction and vibrates downward again, they are both thrown upward above the railing. The safety belts around their waists halt them, pulling them back to the basket. They are slammed down on to the diamond-plate steel deck of the basket. The basket of the aerial platform continues to vibrate up and down, and the aerial stream jets out at random, uncontrollably.

The building lies in a smoldering pile beneath the aerial platform boom.

Officer and firefighter scramble up from the basket floor. The firefighter grabs the controls of the aerial stream and directs it downward into the smoldering, collapsed building. The captain grasps the operating controls.

"Are you okay?" the officer asks the firefighter.

"Yes, I think so. My hand hurts. I might have broken something."

The captain looks down to the street.

"I guess they forgot about us up here. Ladder 6 to turntable control. Shut down the supply line to the aerial stream. The ground handlines can take over now."

"Ten-four, Captain," turntable control returns. "Boy, that was some collapse—are you guys okay up there?"

Figure 13.1. *Most firefighters are killed by collapsing walls outside burning buildings.*

Firefighters operating outside burning buildings must know how to protect themselves against injury from the collapse of those buildings.

Most firefighters killed by falling walls are outside the burning building (Figures 13.1, 13.2, 13.3, and 13.4). Firefighters must be trained in defensive outside fireground tactics. For example:

- Firefighters must understand the concept of the collapse zone.
- Firefighters must know how to use the reach of a hose stream as a safety device.
- Firefighters must know the safety advantages of using an aerial stream.
- Firefighters must know how to calculate the collapse danger zone for an aerial stream.

Figure 13.2. ABOVE LEFT: *A collapsing wall can fall three ways: at a 90-degree angle, in an inward-outward configuration, and in a curtain-fall collapse.*

Figure 13.3. RIGHT: *A 90-degree angle is the most dangerous type of collapse because it creates the largest collapse danger zone.*

Figure 13.4. *A collapse danger zone is that distance away from the unstable wall equal to the height of the wall.*

- Firefighters must know how to outflank a wall in danger of collapse.
- Firefighters must know how to position an aerial ladder or elevated platform truck safely when there is danger of wall collapse.
- Firefighters must know that fireground dangers are just as deadly outside burning buildings as they are inside.

Collapse Danger Zone

A wall of a burning building can collapse in one of three ways: a 90-degree-angle collapse (Figure 13.5), an inward-outward collapse (Figure 13.6), or a curtain-fall collapse (Figure 13.7)

The 90-degree-angle wall collapse is the most dangerous because it creates the largest collapse danger zone. In this case, the wall falls outward from the building a greater distance than it would in either an inward-outward collapse or a curtain-fall collapse. A wall that collapses at a 90-degree angle will fall straight outward, in a manner similar to a falling tree cut by a woodsman. It will cover the ground below with deadly bricks and timber for a distance at least equal to the height of the falling wall section.

If 25-foot-high wall collapses at a 90-degree angle, it will fall straight outward for 25 feet. No one within 25 feet of the building will survive; all will be buried beneath the collapsed wall.

Whenever there is a danger of wall collapse, plan for the worst—a 90-degree-angle collapse.

Establish a collapse danger zone equal to the height of the wall. Withdraw all firefighters from the burning building for a distance at least equal to the height of the wall—the so-called "vertical" or "outward collapse" danger zone.

Next, estimate how much of the wall's horizontal area may collapse. This area is the so-called "horizontal" collapse danger zone (Figure 13.8). For example, consider a 25-foot-high wall 100 feet long, with a parapet wall, over a row of stores, and a ten-foot-long bulge at the center. The horizontal collapse danger zone may be a horizontal area ten feet directly in front of the bulge; however, a firefighter or fire officer must know whether or not the walls are tied together by steel reinforcement rods. If they are so tied together, a small portion of unstable wall could pull down the wall's entire horizontal length. In this instance, it may be necessary to establish a horizontal collapse danger zone equal to the entire wall frontage or building width.

90-Degree-Angle Wall Collapse

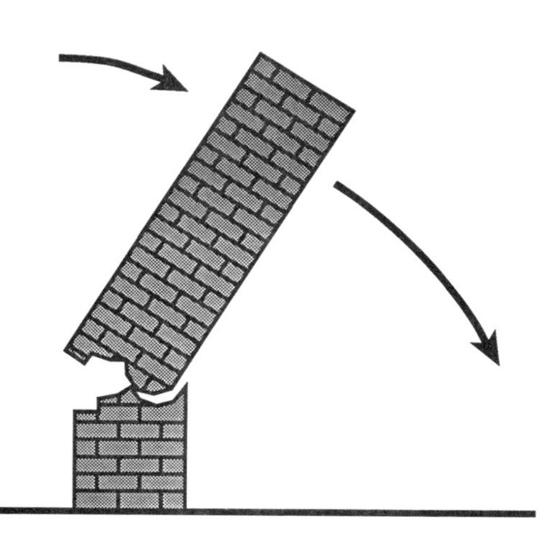

Inward-Outward Wall Collapse

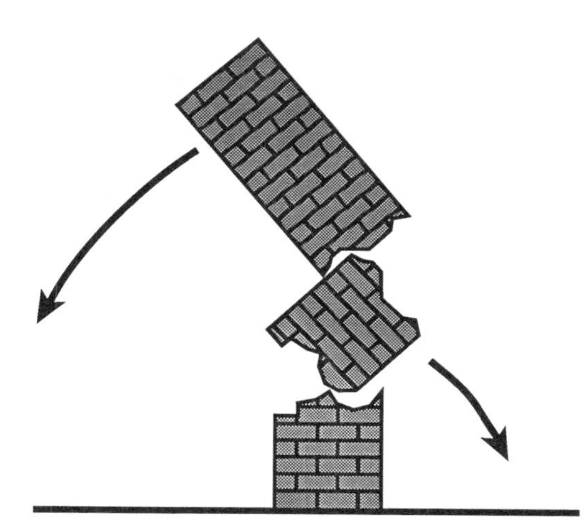

Curtain-Fall Wall Collapse

Figure 13.5. ABOVE LEFT: *A wall falling at a 90-degree angle collapses straight outward, in a manner similar to a falling tree.*

Figure 13.6. RIGHT: *An inward-outward collapsing wall breaks into two pieces. One part falls back into the building, and one part falls out, away from the building.*

Figure 13.7. *A curtain-fall collapsing wall falls straight downward like a curtain cut loose at the top.*

Collapse Danger Zones – 2

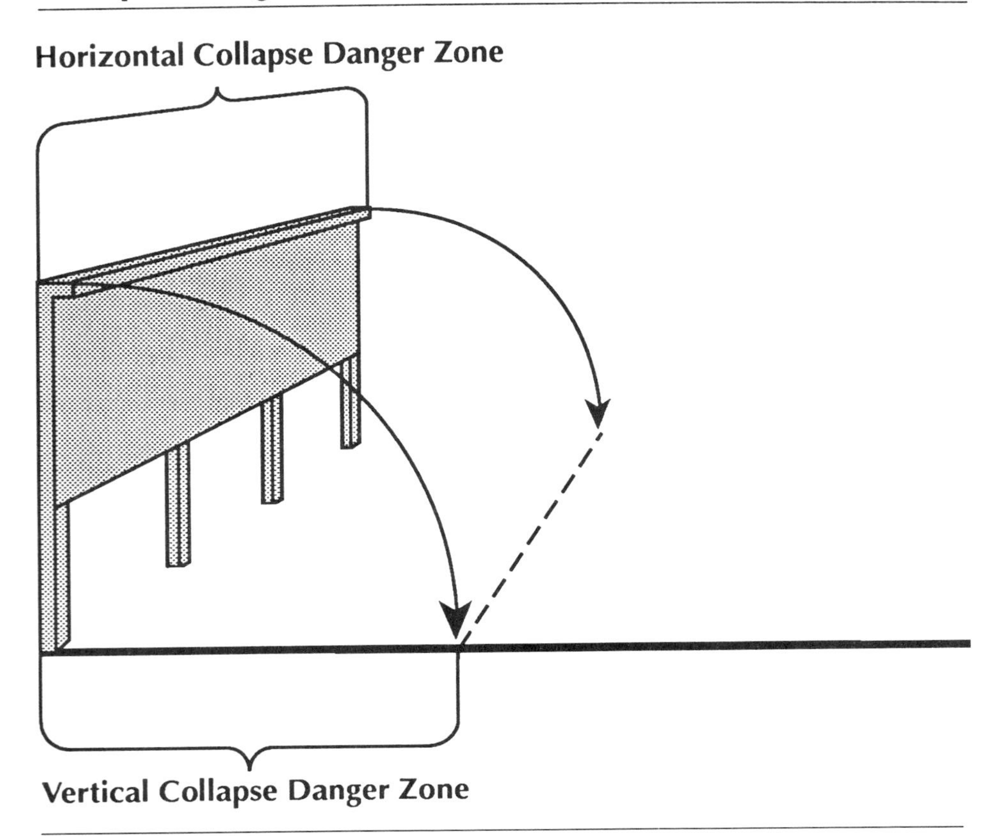

Figure 13.8. *A horizontal collapse danger zone may be equal to the width of the entire wall frontage or building.*

Reach of Hose Stream

When a collapse danger zone is established, the fire officer is trading off the advantage of close-up hose stream penetration through an open door or window of a burning building for the greater advantage of firefighters' safety. This practice conforms to the commandments of firefighting: life safety first—and this rule includes the life safety of firefighters as well as that of civilians; fire containment second; and property protection third.

The reach of a hose stream is then used as a safety measure. It allows firefighters to move away from a dangerous wall and still discharge water on the fire. The typical reach of a hand-held hose stream is 50 feet; that of a master stream is 100 feet. By utilizing this distance as a safety device, a 50-foot hose stream could allow fire-

Using the Reach of a Hose Stream to Increase Safety on the Fireground

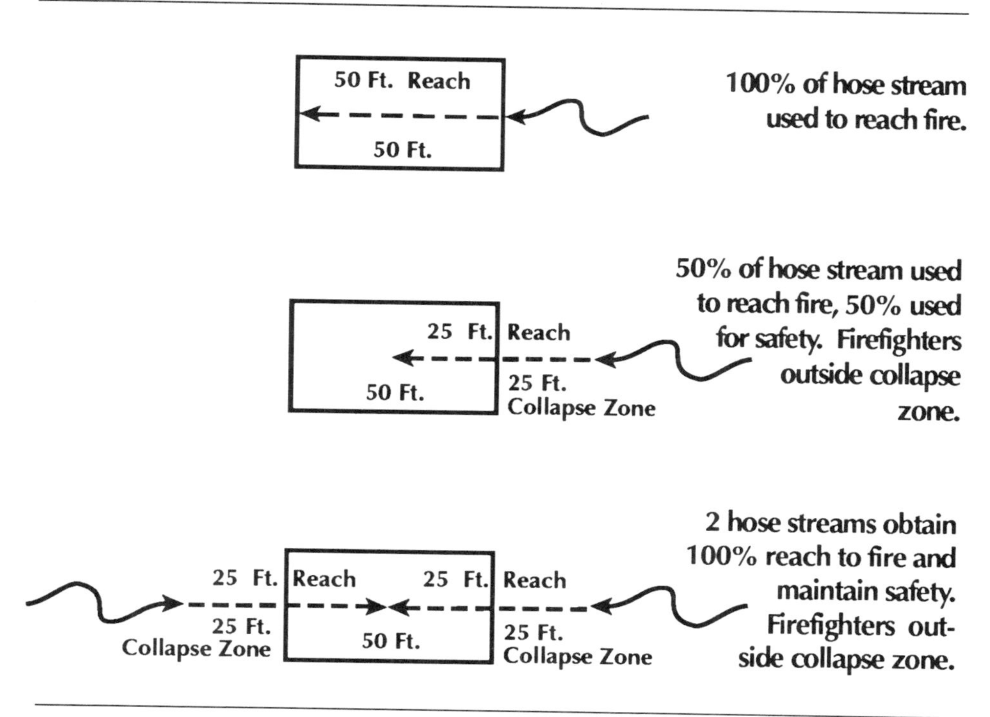

Figure 13.9. *Three positions from which to operate master streams near a wall in danger of collapse are: An* aerial stream *can safely discharge over the top of a wall; a* portable deluge nozzle *can be positioned in a flanking position in front of an adjoining building; and a* deck gun *atop a pumper can be safely operated outside the collapse danger zone.*

fighters to operate outside a collapse danger zone. The officer could withdraw firefighters 25 feet away from a dangerous wall and still discharge water into a burning building from a distance of 25 feet (Figure 13.9). In increasing fireground safety, maximum hose stream effectiveness is thus sacrificed. Yet another hose stream could be positioned at another side or at the rear of the burning building in order to regain fire stream effectiveness. Such a fireground strategy increases fireground safety.

Aerial Stream Use

The most effective use of an aerial stream from a ladder, aerial platform, or snorkel takes place from below the roof through a window or other wall opening in a close-up approach to the burning building. Even when firefighters are battling a fire in a one-story building, an aerial platform may be used at a low angle, close to the ground (Figure 13.10). When large, plate-glass windows have been removed by flames or by firefighters for ventilation purposes, and when there are no interior partitions or walls, a powerful stream from an aerial platform can sometimes sweep the open space and quickly extinguish a large fire. When there is danger of wall collapse or where interior partitions or stock subdivide a row of small stores, however, and fire burns through the roof, it's necessary and safer to operate the stream from above. This method is less effective, but it becomes a great advantage in maintaining the safety of firefighters operating in the basket or at the tip of the aerial when there is danger of building collapse (Figure 13.11). Moreover, when interior partitions and/or stock block stream penetration during an outside attack, there is no alternative but to operate from above. When firefighters are operating an aerial stream from above a dangerous wall, directing the stream downward at a fire burning through a roof—the height of the aerial is the safety device. The wall may collapse outward at a 90-degree angle, but the firefighters in the basket or at the tip of the aerial are out of the collapse danger zone; they are above it.

Another safety tactic for aerial operations is to operate the aerial ladder without firefighters at the top rungs (Figure 13.12). The top of an aerial can be extended outward from its base, the centerline of the turntable approximately 30 feet from the structure. When the aerial stream is directed with halyards by a firefighter on the turntable, the ladder's extension and height provide an extra 30 feet of reach. This extra 30-foot reach can be added to the 100-foot reach of an aerial master stream. Firefighters could withdraw 65 feet away from a dangerous wall and still direct water to penetrate into a burning building 65 feet high.

Collapse Danger Zone for Aerial Streams

When a fire occurs in a multi-story building and there is danger of wall collapse, the chief and the fire officer sometimes overestimate the capabilities of the aerial stream and underestimate the dangers to the firefighters operating at the top of the aerial device. In some instances, we withdraw the ground streams away from a potential explosion or wall collapse danger; then we maneuver the top of the aerial platform or aerial ladder close to an unstable wall in order to extinguish fire. The aerials thus enter the same collapse danger zone from which the ground forces were withdrawn.

Figure 13.10. ABOVE LEFT: *When there is no danger of wall collapse, an aerial platform or ladder pipe may be used at a low angle, close to the building wall.*

Figure 13.11. ABOVE RIGHT: *An aerial stream, operated from above, hitting fire that is burning through a roof is less effective from a fire-extinguishment point of view; however, it provides safety for firefighters assigned to operate in the basket of the aerial platform.*

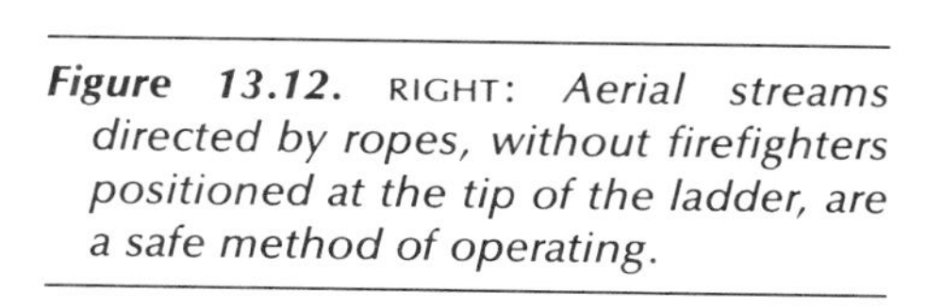

Figure 13.12. RIGHT: *Aerial streams directed by ropes, without firefighters positioned at the tip of the ladder, are a safe method of operating.*

Figure 13.13. An aerial platform with its bucket torn from the boom, after a building wall collapsed on it. Collapse danger zones should be set up for aerial streams as well as for ground streams.

In recent years, an increasing number of building collapses have killed and seriously injured firefighters operating master streams at the top of aerial devices. An officer and two firefighters were crushed in the bucket of an aerial platform when one wall of a three-story wood-frame building suddenly collapsed. In another incident, an aerial platform had its basket torn from the boom when a three-story, wood-frame building collapsed on it during an outside attack on a fire (Figure 13.13); two firefighters in the basket were thrown 30 feet down into the burning collapse rubble and were seriously injured.

The collapse danger zone for an aerial stream will vary from the collapse danger zone for a ground stream. The apparatus must be positioned in the street outside the collapse danger zone, and the raised aerial ladder or platform should never enter inside the arcing path of the unstable wall, which could collapse at a 90-degree angle (Figure 13.14).

Collapse Danger Zone Arc

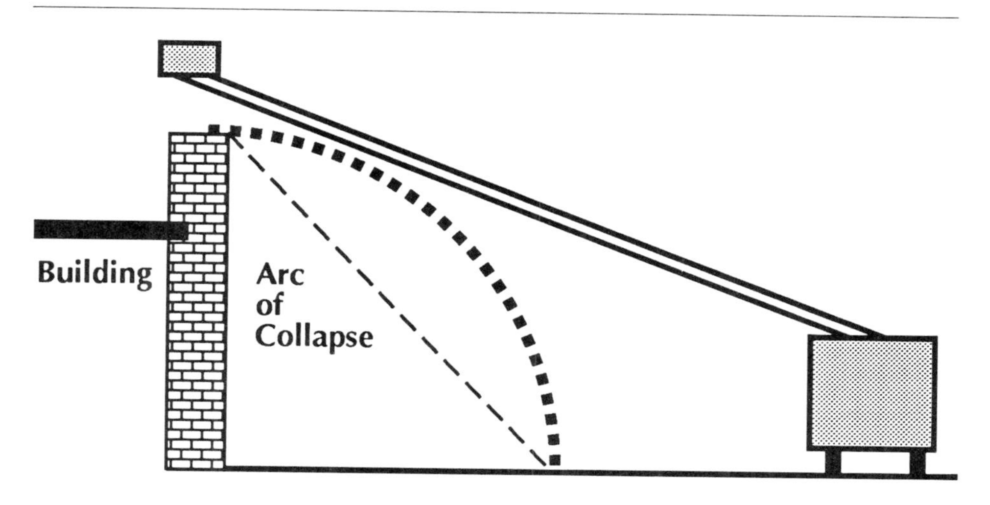

When an aerial stream is operating and there is no danger of collapse, the tip of the aerial device may be placed as close to the building as possible. This will provide deep stream penetration at the proper angle.

When there is a danger of wall collapse, however, the aerial tip, boom, and truck should be outside the collapse danger zone. This includes the arc of a wall falling at a 90-degree-angle collapse.

Figure 13.14. *Firefighters operating at the tips of aerial ladders or aerial platforms must consider the arcing curve of a collapsing wall, falling at a 90-degree angle.*

Flanking a Fire

When a tall structure is involved with fire and there is danger of a wall collapsing, establishment of a collapse zone could require firefighters to be positioned so far away from the unstable wall that the hose streams do not reach the fire.

During a fire inside a church that is 80 to 100 feet in height, for example, a collapse zone would require ground and aerial streams to be operated 80 to 100 feet away in order to avoid falling walls. In this instance, a fire officer may position ground streams in a "flanking" position; that is, firefighters must outflank the dangerous wall and continue to direct hose streams.

To accomplish this task, portable deluge nozzles must be placed in front of adjoining buildings or in corner safe areas of the fireground

Three Positions of Master Streams When There Is a Danger of Wall Collapse

1 **Aerial stream from above the wall**
2 **Ground stream outside the collapse danger zone**
3 **Ground stream flanking a dangerous wall**

Figure 13.15. *Master streams may be placed in a flanking position around a door opening when there is a danger of explosion. Master streams may also be placed in a flanking position in front of adjacent buildings when there is a danger that the burning building will collapse.*

(Figure 13.15). The portable deluge stream's range and effectiveness will be reduced, but the life safety of the firefighters will be ensured even if the unstable wall falls outward beyond the collapse danger zone.

Corner Safe Areas

The greatest danger of wall collapse occurs during an outside attack on a fully involved church fire. The side walls usually support the truss roof timbers; when the roof collapses, it will cause the side walls to collapse outward (Figure 13.16). The front wall with its tall steeple can collapse at any moment during the fire; the same holds true for the rear of the church.

When we look at the fireground around a burning church from a bird's-eye view, we see there are only four safe areas in which to park vehicles and operate master streams: the four corner safe areas. If all the walls collapsed outward simultaneously (which is unlikely), only four areas would be safe from falling brick—the corner areas of the square or rectangular burning structure. Master ground streams and vehicles providing aerial streams should be positioned at these four

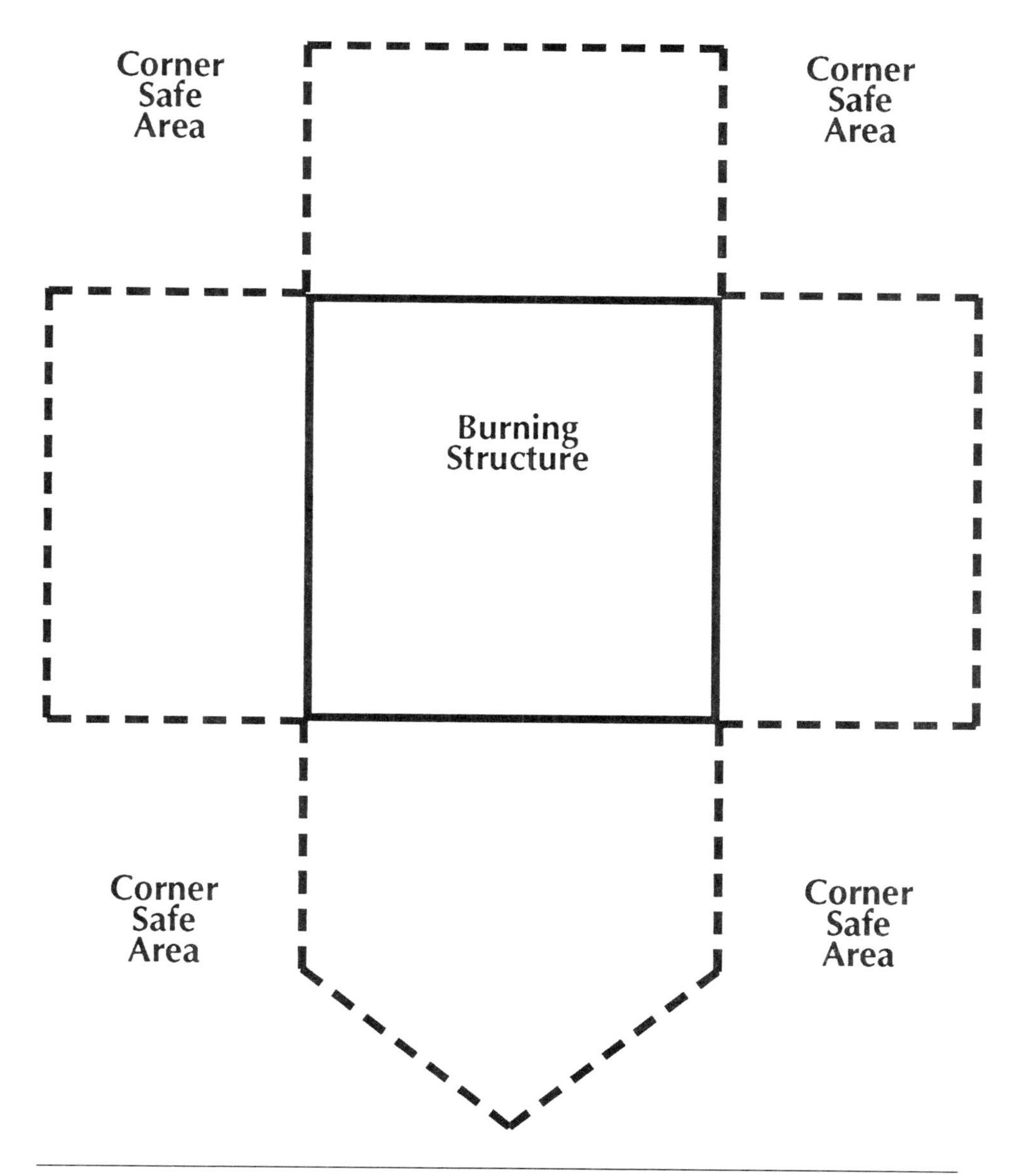

Figure 13.16. *When we look at a burning building from a bird's-eye view, we see that there are four corner safe areas of the fireground, which would not be covered by collapsing walls. Fireground commanders should order apparatus and master streams positioned in these corner safe areas when there is danger of collapse.*

corner areas when smoke conditions and radiated heat make it possible.

Lessons to Be Learned

The most effective method of extinguishing a building fire is by an interior attack. At some fires, however, due to the area of flames encountered upon arrival, the high flammability of contents within a structure, or the combustible nature of the structure itself, an outside attack must be the only strategy. An outside attack at a rapidly spreading fire can be more dangerous to firefighters than an inside firefighting procedure. At these fires, the flames are so large and the radiated heat so intense that the firefighting forces cannot even approach the burning structure. Fire apparatus initially positioned at hydrants, or with ladders raised, has to be hastily repositioned to avoid the radiated heat waves and collapsing walls caused by the fire. At these rare fires, the strategy may be simply to protect exposed buildings and let the original fire building burn itself out.

Firefighting objectives are: life safety, preventing extension, confining the fire, extinguishing the fire, salvage, and overhaul. Remember this order of importance at all times, and know how to employ correct defensive outside firefighting tactics. Your life, and the lives of other firefighters, may depend on it.

14

The Dangers of Outside Venting

"VENT THE REAR of the building!" the officer shouts to the firefighter. "Ten-four, Captain."

The firefighter runs through the front yard of the burning 2½-story frame house carrying a pike pole and a halligan tool. An uncharged 1¾-inch hose line is being stretched to the front entrance. There is no visible flame, but dark brown smoke pushes out of every opening into the early morning daylight. It pours downward from the eaves of the gabled roof, streams from the open attic window, and forces its way out the top of the partly opened front doorway and cracks in the window frames.

Near the entrance in the front yard, a woman in nightclothes clutches a doll and gestures wildly. "My baby! My baby is still inside! Please save my baby!" The firefighter assigned to vent the rear of the building from the outside rushes over to her. "Lady, where inside the house is your kid?" Her words rush out in short gasps: "He's in the bedroom! My baby is sleeping in the bedroom!" Again the firefighter asks, "Where? Lady, where is the bedroom—first floor? Second floor? Front? Rear?" "He's in the crib in the bedroom on the first floor," she cries, "next to the kitchen. Please, I beg you, save my baby!"

The firefighter hurries to the side of the house, pushes open a gate in the small fence enclosing a side yard, and heads for the back of the structure. Halfway there, out of the corner of his eye he sees a small, dark object running toward him: a snarling dog. There's no turning back. Running toward a higher fence in the rear yard, he pitches the tools over it, jumps up, grabs the top railing, and pulls himself over the fence. The dog leaps and snaps at his heels as he escapes.

Once in the backyard, the firefighter sizes up the building. There are two first-floor windows with smoke seeping around their frames. The glass is stained dark brown; condensed smoke is dripping down the inside of the glass panes. Through the left window, he sees flames flickering inside the room. Which window leads to the child's bedroom? the firefighter asks himself. "I've got to get that kid!" He hears shouts of "Start water!" coming from the front of the house.

The firefighter notices a kitchen exhaust fan adjoining the window on his left and concentrates his efforts on the other. Must be the bedroom. The window is old and sealed with paint; he can't force it open. He grabs his pike pole and, standing to the side of the window, swings it at the top pane. *Crash*! Heavy smoke flows out of the top of the window, but no flame. *Crash!* He repeats the procedure for the bottom pane, then hooks the center wood frame with the pike pole and rips it out of the opening.

Dropping the pike pole and picking up the halligan tool, the firefighter looks around. He grabs a metal garbage can, turns it upside down, places it directly below the bedroom window, climbs up, dons his mask facepiece, and slides in over the bottom of the windowsill.

Inside the smoke-filled room, the firefighter stays low. He sees flames through a partly opened door to another room. The smoke is stratified just above his head. He sees a stuffed chair, a table, and a wall in an outer room engulfed in flames. Crawling on the floor, the firefighter shuts the door partially, separating the bedroom he is about to search from the fire in the next room. He hears the hose stream crashing against the plaster walls outside the room. The hose line is knocking down the fire; that's good, he thinks. Smoke and heat inside the bedroom start to lift because of the open window.

Suddenly, the firefighter sees the crib. He feels the mattress and a small lump under the blankets. He picks up the child with both hands, quickly wrapping him in the blankets, and heads back to the window. The small body is motionless.

Laying the bundle down on the floor next to the window, the firefighter quickly climbs halfway out, feet first. Picking up the baby, he jumps first to the garbage can and then to the ground.

Crash! Crash! Crash! Firefighters on the second floor are venting glass windows from the inside. Bending over to shield the small bundle with his back and shoulders, the firefighter runs away from the building, out of the path of the falling glass.

Down on one knee, he frantically pulls the tangled layers of blanket and sheets apart, trying to find the child's face. He finally uncovers a small, soot-covered face with closed eyes. The child is not breathing.

Tilting the baby's head back and holding him with two hands on a bended knee, the firefighter covers the child's small mouth and nose with his mouth and puffs into his small chest. *One* and *two* and *three!* The firefighter feels movement in the blanket—legs and arms kicking and twisting. The baby begins to cry. "Thank God!" the firefighter says.

Standing up, he runs to the front of the house from the other side to avoid the dog. Hugging the bundle, he half kisses, half puffs air into the screaming baby's face.

In the front yard, the mother and the emergency medical personnel tug at the child in the firefighter's arms. The chief, standing nearby, radios to the firefighters inside the burning building who are still frantically searching for the baby, "Urgent! Urgent! All hands inside the fire building. We found the missing baby. The baby is safe," the chief says. "Take it easy—no unnecessary risk taking. Just concentrate on putting out that fire!"

One of the most dangerous firefighting assignments at a fire is outside venting. How can that be, you might ask. Compared to some

Figure 14.1. *Firefighters assigned to outside venting must often work alone during a fire.*

other firefighting jobs, such as advancing an attack hose line and searching above a fire, venting windows seems easy. The firefighter performing ventilation is outside the burning building and all he has to do is break a few windows, right? Wrong! A closer look at outside venting shows that it requires knowledge, skill, and determination.

The firefighter assigned to vent the outside of the building must often work alone, sometimes at the rear or the side of a burning building (Figure 14.1). If he is trapped or injured, no one will see and come to the rescue.

The duties of the outside vent firefighter are varied and dangerous. Just to come close to a window that needs to be vented, the firefighter may have to climb over a fence, fight off an attack dog, force open the lock on a gate, climb over or cut barbed wire, raise and climb a ladder, scramble up on to a porch roof, and lower the counterbalance stairway or access ladder of a fire escape (Figure 14.2).

The following sections present some of the fireground dangers and safe operating techniques a firefighter assigned to outside venting should know about in order to increase his own safety.

Dangers Around the Perimeter

Firefighters venting windows from outside a burning building have been seriously injured and killed by objects falling out of or collapsing off the structure. These firefighters must realize that they are operating in one of the most dangerous areas of the fireground: the perimeter around the burning building. Firefighters inside the fire building are concerned with searching, extinguishing, and venting. They don't always know the exact location of the firefighter venting from the outside because his position varies from fire to fire. As windows are vented from the inside, glass falls out. Hose streams that are extinguishing fire also blast windows outward (Figure 14.3). The outside venting firefighter must be aware of these dangers and try to avoid them, while performing his assigned duties, by positioning himself away from any possible direct path that breaking glass may take.

Falling objects that often injure outside venting firefighters include broken window glass from inside venting, tools that slip from the hands of firefighters operating above, large smoldering pieces of furniture pushed out of the window at night after a quick fire knockdown, bricks or capstones loosened by a hose stream striking a parapet wall or chimney, slate shingles on a sloping peaked roof heated by

Figure 14.2. ABOVE: *Outside venting firefighters must have the skills necessary to perform forcible-entry operations, ladder raising and placement, search and rescue, and window venting.*

Figure 14.3. *Hose streams used for interior attack can blast a window outward into the face of a firefighter performing outside venting.*

an attic fire, and even people who jump from buildings to escape flames (Figure 14.4).

There are several operating precautions firefighters can take to avoid injury from falling objects during outside venting. Eyeshields and a properly fitting helmet are two of them. In addition, all new firefighters are warned in basic training school: "When you hear glass breaking, don't look up." That's sound advice (Figure 14.5).

Size up the venting assignment from a distance, if possible, to avoid falling objects. Choose the window you want to vent, move in close, vent it, and back away from the structure. If, however, the yard or alley is narrow and you can't stay outside the collapse danger zone, stay very close to the building, hugging the wall. Objects falling from upper floors may be propelled several feet away from the building.

There are often areas around a building beneath which the outside vent firefighters may take refuge. A recessed doorway, a canopy over a door, a porch roof, a fire escape balcony, and a decorative overhang or cornice may all provide a temporary protection against falling objects.

Portable Ground Ladders

When a fire occurs on the upper floor of a two- or three-story residence, the firefighter venting outside must place a ground ladder at the rear or side of the building in order to reach a window. He may have to position the ground ladder up to the roof of a porch, to a one-story extension, or to the window itself. Outside venting firefighters climbing ground ladders can be injured in two ways: by placing the base of the ladder at a precarious angle, from which it slides out under the climbing firefighter (Figure 14.6), and he falls from the ladder, or by falling off the tip of the ladder, because the window explodes outward into his face as he is about to vent it.

To ensure that the base of a ground ladder does not slide out from under you, place the ladder at a proper climbing angle. The base of the ladder must be at an optimum distance away from the wall of the building. There are many formulas for calculating this distance (ladder length divided by four or ladder length divided by five plus two); however, during a fire there is usually no time to calculate. Here's the quickest and best way to determine if the ladder is placed at a safe angle. After raising the ladder, stand erect at the base of it with your boots against the beams of the ladder and your arms stretched forward, grasping the rungs at shoulder level. If you can do this, the

Figure 14.4. ABOVE LEFT: *Falling objects around the perimeter of a burning building can injure a firefighter performing outside venting.*

Figure 14.5. ABOVE RIGHT: *When you hear window glass breaking, don't look up.*

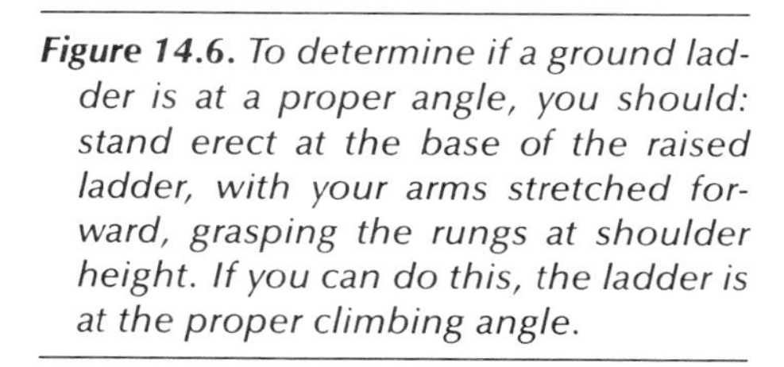

Figure 14.6. *To determine if a ground ladder is at a proper angle, you should: stand erect at the base of the raised ladder, with your arms stretched forward, grasping the rungs at shoulder height. If you can do this, the ladder is at the proper climbing angle.*

ladder is at the proper climbing angle. If not, the ladder should be readjusted.

A ground ladder can still slip, especially if the base is resting on broken pieces of glass strewn over a concrete patio or driveway, if the surface is icy or wet, or if the ladder base moves as a result of the vibrations caused by climbing or operating near the ladder tip. In life-threatening situations, you may have to climb a ground ladder at an unsafe angle. If so, either tie the ladder to an immovable object, have another firefighter hold the ladder in place, or, if possible, quickly create a small hole or depression in the earth to hold the ladder base securely.

An important decision is the place to position the tip of the ground ladder. This choice depends on what you want to accomplish after sizing up the outside of the fire building. If you want to vent as many windows as possible, place the ladder between two windows. If there is only one window to vent and it may explode outward from the force of an interior hose stream, an explosion, or the fire, place the ladder to the side of the window. If your objective is to enter the window and conduct a primary search, place the ladder at or below the windowsill to facilitate victim removal.

Window Removal

Different circumstances warrant different venting procedures. In some instances, the firefighter can quickly force the window open and vent the smoke-filled room without breaking the glass, or he can break the glass window and preserve the window frame. At fires where building entry or victim removal is necessary, he may have to remove the glass, and the entire window frame may also have to be removed in order to provide access in and out of the smoke-filled room (Figure 14.7). The firefighter may be ordered to vent an unoccupied building solely for purposes of removing smoke and heat, thereby assisting the advance of the interior attack team; in that case, a six- or eight-foot pike pole provides greater reach and safety. It allows him to operate farther away from the window, reducing the risk of being blasted by scalding water from interior hose streams or of being hurled off the ladder by an explosion.

Cuts and other wounds to the skin comprise the second leading cause of fireground injury (the leading cause in strains and sprains). The outside vent firefighter often receives glass cuts and wounds from the windows he vents, those vented above him, and those vented from the interior (Figure 14.8). If you cannot open a window manually and

Figure 14.7. TOP: *Remove the entire window frame when entering a window to perform search and rescue. This precaution will assist in victim removal or allow a rapid escape.*

Figure 14.8. *Cuts and wounds to the skin comprise the second leading cause of fireground injury.*

Figure 14.9. *To vent a window safely with a pike pole when operating from a ground ladder, strike the top of the window first; then work downward. Position the ground ladder to one side of the window to avoid falling glass when you are venting it.*

must break it, stand to one side of the opening (the windward side, if possible), strike the glass with the pike pole at the top of the window and then work downward (Figure 14.9). If there is a possibility that firefighters are searching inside the room, first tap on the window and break only a small portion of the glass—this action will serve as a warning. Then remove the entire window with the tool. Keep the eyeshield on your helmet down for protection, wear gloves to protect your hands, and don't stand in front of the glass window you are about to vent.

Cut Off by Flames

Firefighters venting upper-floor windows from the outside must often gain access to porch roofs, fire escapes, or the roofs of one-story extension buildings. Portable ladders provide access to these roofs or balconies—but when fire breaks out of a window and cuts off the outside venting firefighter from the portable ladder, he may be trapped on the roof or balcony (Figure 14.10). Such instances are now occurring more frequently. At many fires, combustible liquids used by an arsonist, plastic furnishings, or air movements inside the burning structure create and send tremendous convection currents of flame out of open windows.

A firefighter so trapped does have a few options: he can wait until the fire is extinguished and then get safely back to the ladder, call for an additional ladder to rescue him, or perform some acrobatic feat to climb down from the porch roof or the fire escape balcony. In any case, his entrapment disrupts the fire operation; entry and search from that point will have to be suspended temporarily; the inability to

Figure 14.10. *Firefighters venting windows from a fire escape or a porch roof can be trapped by rapidly spreading flames.*

Figure 14.11. When operating from a porch roof or a fire escape, vent windows farthest from the ladder first; then work back to the escape ladder.

perform the venting tactic could impede the progress of the interior attack team; his intrapment prevents him from performing other duties; or it diverts his rescuers away from other important tasks.

Take precautionary steps to prevent such entrapment. When you need access to a porch or a one-story extension's roof, place the portable ladder on the windward side, if possible. When several windows require venting from the roof, vent the window farthest from the ladder first, working back to the window closest to the ladder (Figure 14.11).

Venting windows from a fire escape is more difficult. First, the ladder giving access to the fire escape balcony is in a fixed position. It may be on the more dangerous, leeward side of the windows about to be vented. In addition, there is less room to work on a fire escape balcony, compared with a one-story extension's roof or a porch roof.

Firefighters must stay low, crouched down, so flames and heat from a self-venting window can pass over them. Vent windows farthest from the fire escape ladder first, working back to the ladder when there is a serious fire and a danger of flame emerging from the window.

Sometimes the fire escape serves a large burning room that has two windows, only one of which is served by the fire escape. Vent the window not served by the fire escape first. Lean over the balcony railing, and use a tool that will afford you some reach. If flames explode out of that window, they will relieve some of the heat and pressure in the room that might, otherwise, self-vent the window served by the fire escape.

Outside Entry and Search

Outside venting becomes more dangerous when there is a report of a trapped person inside the fire building. The venting firefighter still must locate the window to be vented—the one located at the side of the fire opposite from that which the hose attack team advances. Only now he must vent and also make an attempt to search the vented room, and possibly other rooms behind the fire, for trapped victims, if fire conditions permit.

After venting a window, if flames and smoke explode from the upper portion of the window opening or if the room appears to be about to flash over, do not enter the flaming room; however, do not simply proceed to another assignment, either. Instead, crouch down low below the flames and attempt to sweep the area directly beneath the windowsill inside the room with a tool or with your gloved hand (Figure 14.12). Sometimes you will find people alive, slumped over

Figure 14.12. *Sweep the area beneath the windowsill inside a room with a tool or your gloved hand.*

Figure 14.13. When unable to search burning rooms, search adjoining rooms until the blaze is extinguished; then complete the primary search.

each other, directly below the sill inside the flaming room. They collapse there after they lack enough strength to open the window or after calling for help from the partially opened window. If you are unable to enter after sweeping the area, try venting, entering, and searching adjoining rooms. Then, after the blaze is extinguished, you can enter the room and complete thorough primary and secondary searches (Figure 14.13).

In some cases, the flames are not in the room where you vented the windows but rather in an interior room. In this situation, the outside vent firefighter may enter the vented window, proceed to a door separating the fire from the front rooms, and close that door. The closed door thus acts as a flame barrier. The firefighter can now search the rooms and remove any victims. If there are no trapped or unconscious victims, the firefighter can, if fire conditions permit, reopen the door he closed before the search to provide a path of ventilation. He must then retreat quickly by way of the vented window.

Flashover

Flashover is a stage of fire growth in which the smoke-filled, superheated room suddenly bursts into flames. This abrupt change can trap and kill outside venting firefighters who enter rooms to search. There are two warning signs of flashover: a build-up of heat inside a smoke-filled room, and "rollover"—flashes of intermittent fire mixed with smoke, either inside a room or flowing out a door or window. When a firefighter enters a smoke-filled room and finds very little heat mixed with the smoke, flashover is unlikely. If, however, you are searching a smoke-filled room and have to crouch down to the lower half of the room's height in order to get below the superheated gases banked down below the ceiling, that room is definitely in danger of flashing over.

In some instances, during a prolonged search inside a smoke-filled room, there is little heat; however, after a period of time, heat builds up above the head of a firefighter who is crouched down below the smoke to increase visibility during his search. During an extended search in a smoke-filled room, raise an ungloved hand into the smoke above your head to check for heat build-up. If the intensity of heat prevents you from raising your hand, put your glove back on and get out. Flashover could be imminent.

Rollover precedes flashover. When intermittent flashes of flame mix with smoke at ceiling levels, flashover is about to occur. Unfortunately, because the first signs of rollover occur at the upper levels of the smoke-filled room, the outside vent firefighter crawling around the room searching for trapped victims cannot see them. Therefore the build-up of heat is the only sign of flashover that a firefighter inside the smoke-filled room can detect.

Outside, however, you can see rollover directly outside the window or door just opened for ventilation. If smoke flows out a window just vented at its upper levels and you can see flashes of flame where the convection currents of smoke and heat are flowing, flashover can occur. Firefighters should not enter that room to search but should sweep the area below the sill, search an adjoining room through another window, if possible, or enter the burning structure from another point and search the room after the fire has been extinguished.

After flashover occurs inside a superheated, smoke-filled room, there is a point of no return beyond which a firefighter cannot escape back to safety. When a room flashes over into flame, the firefighter is exposed to temperatures of 1,000° to 1,500° F, according to the "time-

Point of No Return

Figure 14.14. The point of no return for a searching firefighter in a room about to flash over is five feet.

temperature" fire curve. Tests show that exposure to temperatures of only 280° to 320° F will cause extreme pain to all unprotected skin. The average speed of a crawling firefighter with 60 pounds of safety equipment is roughly 2.5 feet per second. A firefighter caught in a flashover ten feet inside a room will be exposed to temperatures of 1,000° to 1,500° F at the neck area for approximately four seconds before he can escape out the window or door—that is, providing he can find either one. Thus, the point of no return, or the maximum distance a firefighter can crawl inside a superheated, smoke-filled room and still escape alive after flashover occurs is only five feet (Figure 14.14).

Lessons to Be Learned

Outside venting is an extremely dangerous task. It should only be given to the experienced, knowledgeable firefighter who understands the priorities of firefighting risk-taking—that protection of life comes first. This category includes the life of the outside vent firefighter as well as that of the trapped civilian (Figure 14.15). A firefighter only risks his life to save another life. He should make every effort to rescue a trapped person inside a burning building, short of becoming trapped himself. When fire conditions are severe and there is only a report of a trapped person, but he hears and sees no one, his is the number-one life priority.

Dangers of Outside Venting

1	**Falling objects around the perimeter of a building**
2	**Falls from ladders, porch roofs, fire escapes**
3	**Cuts from broken glass from vented windows**
5	**Flashover during search and rescue operations**
6	**Burns from the blast of a hose stream when operating at the rear of the fire building**

Figure 14.15. *Common injuries suffered by firefighters performing outside venting.*

15

The Dangers of Fire Escapes

"THE CHIEF TOLD US TO 'take-up'," the captain says to the firefighters, "but before we go, let's secure the building. There are a couple of broken windows that have to be cleared of glass shards, the roof scuttle cover we vented must be put back, and we have to raise and reset the fire escape drop ladder."

"I'll get the drop ladder," one of the firefighters responds, as he climbs out of the window of the two-story brick row house. Standing on the lowest balcony of the fire escape, the firefighter temporarily hangs his pike pole over the railing. With both hands free, he reaches down, grabs the metal rung of the vertical drop ladder resting on the sidewalk, and pulls it upward. Hand under hand, he pulls each rung up waist-high. When the ladder is all the way up, the firefighter holds the heavy iron ladder with one hand and swings the pendulum hook under the nearest drop ladder rung to secure it with his other hand.

Instead of reentering the window of the burned-out apartment and walking down the stairway to the apparatus, the firefighter decides to take a shortcut. He drops his tool to the ground and climbs over the fire escape railing and on to the raised drop ladder. I'll climb down the iron ladder as far as it goes, hang from the lowest rung, and jump down the rest of the way, he thinks to himself.

When the full weight of his body is on the ladder and the firefighter starts to climb down, the iron rung holding the pendulum hook breaks, causing the hook to release the drop ladder. The ladder with the firefighter standing on it shoots straight downward, crashing into the sidewalk. The sudden stop knocks the firefighter backward to the ground, flat on his back. Two firefighters run over to help him.

Writhing in pain on the sidewalk, the firefighter cries out, "My feet! My feet! I'm really hurt! Don't try to pick me up! Get an ambulance!"

The next day, word spread quickly throughout the firehouse about the fire escape drop ladder giving way with Bob on it—all the bones in his feet were broken.

At the next fire to which you respond, the chief could order you to stretch a hose line up a fire escape. Your community may not have buildings with fire escapes; however, an adjoining town with which your fire department has a mutual-aid agreement could have many old, rusted fire escapes hidden from view in backyards and side alleys.

Fire escapes possess valuable firefighting advantages in addition to their original purpose of providing a secondary exit for the occupants of a multi-story building. Firefighters can use fire escapes as a means of access to open a window instead of forcing a door, or as a platform from which to vent a window in the path of an advancing interior hose line attack. They can launch a search or rescue mission from a fire escape or use it to make a hose line advance when the interior attack fails (Figures 15.1, 15.2, and 15.3). There are dangers

Figure 15.1. *Fire escapes provide a platform from which a firefighter can launch a search and rescue mission into a burning building.*

Figure 15.2. *Firefighters doing outside venting can use fire escapes to vent windows in order to release pent-up smoke and fire gases.*

Figure 15.3. *Hose lines can be advanced from fire escapes.*

Figure 15.4. Fire escapes assist firefighting operations in many ways.

associated with fire escapes, but there also are safe operating procedures which you can use to minimize those dangers (Figure 15.4).

Types and Functions of Fire Escapes

In the suburbs and in some rural communities, fire escapes are attached to 2½-story dwellings. When an attic is occupied, a fire escape is often attached to the outside of the structure near the attic window in order to provide a second exit from the attic (Figure 15.5). Multi-story public schools, nursing homes, movie theaters, hotels, and apartment houses also have fire escapes.

There are three types of fire escapes; some are more dangerous than others. The exterior screened stairway is the safest (Figure 15.6), followed by the party balcony, and then by the standard fire escape, which is the most hazardous because of its mechanical parts and narrow, high-pitched stairway. The older or more corroded a fire escape, the more dangerous it is. A rusted standard fire escape on the side of a 50-year-old abandoned building is the most hazardous fire escape a firefighter can encounter. If possible, use a fire department aerial or a ground ladder instead.

Figure 15.5. *A standard fire escape is often attached to the attic of a private dwelling.*
Figure 15.6. RIGHT: *An exterior screened stairway is often found attached to a school building.*

EXTERIOR SCREENED STAIRWAY

This type of fire escape is enclosed by a shoulder-high metal screen or railing. The stairway is similar to an interior stair except that it is made of metal and is outside the building. The width and angle are the same as those of a building stairway—two people can descend side by side. The dimensions of step rise and tread are the same as those of an inside stair. The fire escape may have a metal covering over its top to protect users from rain or snow. Its most important feature is that access from the lowest balcony to the street is by way of a permanent metal stairway—there are no mechanical ladders or counterbalance stairways supported by cables and heavy metal weights that must be activated, as is the case with standard fire escapes.

Figure 15.7. *Party balconies have no stairs or ladders connecting the balconies. The escape route from a party balcony is by way of the adjoining occupancy.*

THE PARTY BALCONY

Unlike the other types, the party balcony has no stairs or ladders connecting fire escape balconies serving each floor (Figure 15.7); there is no access to balconies above or below or to the street. This type is strictly a horizontal emergency exit on the outside of the building; it provides an escape route by way of the adjoining occupancy. A person fleeing a fire opens a door or window, climbs out to the outside party balcony, walks several feet to the adjoining occupancy, and enters it through a door or window. The real protection it provides is the unpierced fire division between the two adjoining occupancies.

STANDARD FIRE ESCAPE

This structure is the most common type of fire escape found on residential buildings. It consists of a series of metal balconies interconnected by a narrow stairway or ladders. The top balcony may have a gooseneck ladder leading to the roof of the building, and the lowest balcony has a sliding drop ladder or counterbalance stairs providing access to street level. There may be a vertical ladder connecting balconies on this type of fire escape or a narrow stairway at a steep 60- or 75-degree angle that only one person can use at a time. The step has a rise greater than that of a normal interior stair and a tread smaller than that of a stairway. Only one knee-high thin bar may serve as a handrail along the stairway. The balcony is enclosed only by a waist-high railing.

Protection Against Hazards

The fire escapes found on most old buildings are very different from the sturdy, freshly painted fire escapes found on the drill towers of training centers. Some have been attached to walls of buildings for half a century or more and have become extremely dangerous to use because of neglect or lack of maintenance. The following are some hazards and safe operating procedures for climbing or operating on fire escapes during fires or emergencies.

Climbing

When a firefighter climbs up or down a weakened fire escape, the impact of his body weight pounding on a corroded metal step can make the step collapse suddenly and cause him to fall. The firefighter may tumble down the fire escape stair holding sharp tools, topple over the fire escape railing and fall to the ground below, or catch his leg between the steps and break a kneecap.

The most common injury to firefighters on a fire escape is caused by step collapse. Corrosion weakens the connection between the step and the side stringer to which it is attached. This small space between the step and the stringer, where the connecting bolt or weld is located, is inaccessible to such normal maintenance procedures as scraping and painting (Figure 15.8). In many instances, visual inspection does not detect such weakness.

To protect yourself from step failure while operating on a fire escape, remain aware of the danger at all times and climb each step as if it were about to collapse. Continuously grip some portion of the fire escape with your hand; if the step fails, this precaution will prevent you from falling. Press down on each step smoothly, if possible—do not stamp your foot down as you climb or descend. Placing one foot on the step above or below, apply pressure to it before trusting your full weight to it. Finally, place your foot close to the side of the fire escape step. This position causes less deflection of the metal step, thus

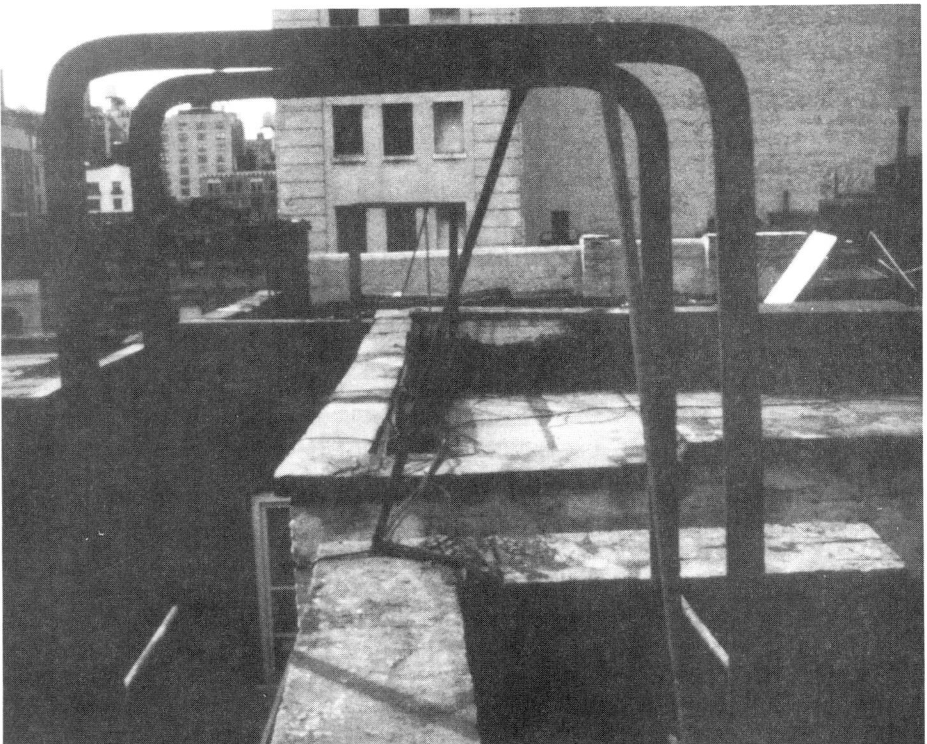

Figure 15.8. TOP: *The most common injury suffered by firefighters on a fire escape is caused by fire escape step collapse.*

Figure 15.9. *A sudden movement of a gooseneck ladder could cause a climbing firefighter to lose his grip and fall backward, off the fire escape.*

reducing the stress on the connections between the tread and the stringer.

If you discover one broken step on a fire escape, it indicates that the other steps are weak and may collapse beneath you. If you suspect that a step may collapse, descend the fire escape backward, facing the stairs. If a step suddenly breaks, you will fall into the stair, not outward—and possibly over the balcony railing.

Gooseneck Ladders

Some standard fire escapes on flat-roofed buildings have vertical ladders that extend from the top-floor balcony to the roof level; the top of such a vertical ladder is curved as it extends over the roof, hence the name "gooseneck ladder." This type of ladder is attached to the exterior wall of the building and to the roof deck. Sometimes these connections are loose, missing, or corroded, and the ladder moves outward from the building during a climb (Figure 15.9). This sudden movement could cause a firefighter to lose his grip and fall backward off the fire escape into the backyard.

To protect yourself from this danger, before climbing a fire escape's gooseneck ladder you should test its fastening by pulling the ladder away from the wall or roof deck to which it is attached. If it does not budge or pull away from the wall, climb it carefully; if it does pull away, do not climb it. Notify your officer in command so that he may warn other firefighters of this hazard.

Counterbalance Stairways

Found on standard fire escapes, counterbalance stairways are designed to provide access between the lowest fire escape balcony and street level. They are extremely dangerous to operate. Some of these heavy metal structures have not been tested or operated for many years and can collapse when activated. Supported on a pivot, these stairways are balanced in a horizontal position by heavy, cast-iron counterbalancing weights. Several hundred pounds of metal are either attached to one end of the stairway or held up by a pulley and steel cable against the side of the building.

A simple bar prevents the ladder from descending; moving that bar from beneath the stairway activates the counterbalance. A firefighter with a pike pole can do so from street level, or a person using the fire escape can do it manually from the balcony. When the counterbalance stairway starts to descend and the cast-iron balancing weights start to rise, watch out: any moving part of the fire escape can collapse and strike a person standing in the vicinity. There have been incidents in which the entire metal stairway has collapsed on to the sidewalk; the heavy metal weights holding up the fire escape have fallen off the counterbalance ladder; the steel cable holding the cast-iron weights has snapped, becoming a deadly whip; and the entire pulley assembly holding the cable and weights attached to the side of the building has pulled away from the building wall and dropped to the street (Figure 15.10).

When you arrive at a fire where people are awaiting rescue on the lowest balcony of a standard fire escape with a counterbalance stairway, use a department ladder instead of the counterbalance stairway to reach them (Figure 15.11). Climb up the ladder, calm the waiting people, and, if possible, take them inside the second-floor window and down the interior stairs. That route is safer for everyone involved.

Fire Escape Drop Ladders

Standard fire escapes sometimes have vertical sliding drop ladders, held in place on the lowest balcony by a pendulum hook. The hook holding the ladder in place is released when a firefighter standing on the sidewalk places the hook of his pike pole beneath the bottom rung of the drop ladder and raises it several inches. The pendulum hook swings away and the weight of the ladder is transferred to the firefighter's pike pole. The ladder drops straight down as the firefighter quickly removes the hook end of the pike pole from beneath the rungs.

If the drop ladder was not reset properly after its last use and is not encased in its sliding tracks or guides, it can, when it is released, fall away from the fire escape and strike the firefighter operating it (Figure 15.12). For this reason, a firefighter lowering a drop ladder should stand beneath the fire escape balcony. If the drop ladder falls out of its tracks, the firefighter will thus be protected.

Firefighters should never climb down a raised drop ladder. The iron rung securing the pendulum could break under the firefighter's weight if it is corroded or mechanically flawed. Climbing down the raised drop ladder and jumping to the ground is a tempting shortcut that must be resisted—the firefighter risks broken bones in doing so.

Figure 15.10. ABOVE: *When a counterbalance stairway strikes the sidewalk during use, it could cause the collapse of the entire stairway itself or the cast iron weights, the wire cable could snap, or the pulley assembly could break and fall.*

Figure 15.11. *Use a ground ladder in place of the counterbalance stair; it is safer.*

Figure 15.12. BELOW: *If a drop ladder is not encased in its sliding tracks or guides, it can fall away from the fire escape when it is lowered.*

Overloading a Fire Escape Balcony

There is a limit to the weight that a fire escape balcony or stairway can support. On many old fire escapes, which have been corroded by weather (Figure 15.13) and weakened by lack of maintenance (Figure 15.14), the load-carrying capability has been dangerously reduced.

During a fire operation, firefighters, in their eagerness to extinguish a fire with a hose line stretched up a fire escape, sometimes crowd on to the fire escape balcony outside the fire room. Some fireground photographs show six or seven firefighters behind a hose line attempting to move into a flaming window from a fire escape balcony (Figure 15.15). This crowding creates an extremely dangerous condition. The overloaded fire escape could suddenly pull away from the building, causing the balcony to collapse and sending it and the firefighters to the street below.

The officer in charge should limit the number of firefighters working on a fire escape balcony or stairway during an operation. When crowding occurs, some firefighters should be ordered to stand by on a lower balcony or should be directed inside a window. There should be no more than three firefighters on one fire balcony or one stairway between balconies. The total weight of three firefighters, each weighing 200 pounds, carrying 50 pounds of breathing apparatus and fire gear, and advancing 100 feet of charged 2½-inch hose, is one-half ton.

Falling Objects

When cleaning his fire gear back in quarters after operating on a fire escape during a serious fire, a firefighter may find small pieces of broken glass or burned wood in the pockets of his turnout coat or hardened splatters of solder or tar on his helmet or coat. Firefighters working on a fire escape below the fire are in a dangerous area—the perimeter of the burning building. Falling objects such as broken glass, small pieces of burning wood, dripping hot metal solder, molten tar, and even bricks from a parapet wall can rain down the side of the burning building from the burning cornice and the firefighting operations above (Figure 15.16). The metal grillwork of the fire escape allows small objects to fall through it, and these can strike the firefighter.

To protect yourself from such falling objects, don't look up; don't reach over the edge of the fire escape unless it is necessary—stay beneath the balcony above you; wear all protective clothing including gloves; and keep close to the building.

Figure 15.13. ABOVE: *Fire escapes corroded by* weather *can collapse under the weight of a firefighter operating on them.*

Figures 15.14 and 15.15. *Fire escapes weakened by corrosion* (LEFT) *due to* lack of maintenance or weakened by age (RIGHT) *can collapse under the weight of a firefighter during search and rescue operations.*

Figure 15.16. *Falling objects such as broken window glass, burning wood cornice pieces, hot metal solder, and molten tar may drop through the fire escape grillwork and injure firefighters climbing the steps.*

Angle of Fire Escape Ladder

A hidden danger in a standard fire escape is the angle of its stairway and ladders. The climbing angles of a gooseneck ladder, drop ladder, and intermediate stair between fire escape balconies are much steeper than that of a normal building stairway. To compensate for these larger angles, exercise greater caution when ascending or descending a fire escape.

A typical stairway in a building rises at a 30- or 45-degree angle from the horizontal floor level. A standard fire escape stairway rises at a 60- or 75-degree angle from the horizontal floor level; the gooseneck and drop ladder rise straight up, at a 90-degree angle.

The SCBA on a firefighter's back changes his center of gravity—there is actually a constant force pulling the firefighter backward (Figure 15.17). He must be conscious of this force at all times during the climb up or down a 90-degree-angle gooseneck or drop ladder. Momentary release of a grip when climbing hand over hand on the rungs of a fire escape drop ladder will cause the firefighter to fall backward.

Figure 15.17. The weight of the SCBA air tank on one's back changes a firefighter's center of gravity. The weight of the air tank can cause a firefighter to fall backward when climbing a steep stair or a ladder on a fire escape.

Walking down an outdoor fire escape stairway at a 75-degree angle is not the same as walking down an interior enclosed 45-degree-angle stairway inside a building; the height of the fire escape step's riser is greater and the width of the step tread is narrower. This difference changes the body motion required to climb and descend a fire escape from that required for an interior stairway. Also, there is less horizontal space on the fire escape stairway available to maintain balance. You must keep a continuous grip on some portion of the ladder or stairway of a fire escape. Carrying tools or hose, the existence of icy or wet conditions, and the presence of glass shards can make operating on a fire escape and maintaining your balance still more difficult during a fire.

Aerial Ladders and Platforms

Aerial equipment has saved more lives—of civilians and firefighters—than any other advance in firefighting technology. If, however, an aerial ladder or platform is not used properly, it can become a deadly piece of heavy equipment.

A moving aerial ladder or tower ladder boom can cause the collapse of a fire escape (Figure 15.18). If the tip of an aerial ladder being lowered near a fire escape slams down on to the railing, it can cause the collapse of the entire fire escape balcony. Aerial platforms that are positioned at parapet wall and roof levels and are suddenly retracted can pull walls down, dumping tons of brick on top of fire escapes. If a tower ladder bumper pad, at the under side of the bucket, hooks on to a fire escape railing, it can pull the fire escape off the front of the building when the boom is moved away from the building.

Treat aerial ladders and platforms as you would cranes and heavy construction equipment; consider firefighters who operate aerial devices "operating engineers" who position cranes and large construction equipment—and give them special training.

When used improperly, aerial ladders and tower ladders can kill firefighters. In addition to causing the collapse of fire escapes, they have knocked down chimneys, made walls collapse, crushed parked automobiles, knocked down utility poles, themselves collapsed when overextended at low angles, and catapulted firefighters off the tips of sudden-moving aerials.

Firefighters operating aerial devices at fires should use extreme caution at all times, but especially near a fire escape. Place the tip of the aerial ladder against the building instead of against the fragile fire escape. When operating above a fire escape or parapet wall, before

Figure 15.18. *Aerial platforms striking a fire escape can cause it to collapse.*

retracting the aerial ladder or platform first raise the aerial sufficiently to clear the railing or coping stone below, and then retract or move it horizontally.

Lessons to Be Learned

Do not be discouraged from using fire escapes during a fire operation if they can assist in an attack. Fire escapes have many advantages during a fire: they allow occupants to escape; on multi-story buildings, they provide a second way to advance a hose line to an upper floor; they provide quick access to a less damaging type of forcible entry through a window; they provide an outside platform for venting smoke and heat through a window for hose line advance; and they provide an alternate avenue to search for and rescue trapped victims.

Keep in mind, however, that climbing, descending, or operating on a fire escape has its risks; yet, as with all firefighting dangers, you can reduce those risks significantly by following the appropriate safety precautions.

16

Overhauling Dangers

"I WANT THAT tin ceiling opened!" the captain says to his firefighters. Three grimy and soot-covered firefighters, inside what remains of a burned-out tavern, begin the difficult task of pulling down an ornamental tin ceiling to search for hidden fire smoldering in the ceiling space above.

With the Z-end of a metal multi-hook, a firefighter punches up the tin ceiling near a seam, creating a split between two sections. Then the hook is placed inside the small opening, catching the edge of the tin ceiling section. As the firefighter pulls down hard on the metal hook with short, sharp motions, one end of the two-foot by four-foot section of tin ceiling falls down from the wood furring strips, opening up the ceiling. Smoke drifts down from the charred roof beams visible through the opening in the ceiling.

"I want the entire ceiling pulled down!" The officer shouts above the clamor of three hooks being punched up into a tin ceiling, above the cracking of splintered wood furring strips and the alarm bells of several self-contained breathing apparatus, indicating that air is running out.

During the extinguishment stage of the fire, the glass front windows were not vented because the fire was quickly knocked down by the hose team. As a result, the inside of the blackened and charred tavern is extremely hot during these overhaul operations. Water still drips from the ceiling decorations. Puddles on the floor create slippery footing, and the temperature is over one hundred degrees.

The firefighters spread out to avoid injury while they pull the tin ceiling. Section by section, razor-sharp pieces of tin, long pointed nails, ceiling dust, and broken pieces of wood furring strips fly through the air inside the demolished tavern. The more ceiling area that is pulled down, the more smoldering and char are discovered.

One of the firefighters working by himself near the front of the tavern pulls down a section of tin near an electrical ceiling fixture, and his metal hook becomes entangled in the broken electric cable. He stops pulling on the entangled hook as a sharp jolt of electric current shoots through his wet gloves and into his arms. He hangs onto the charged metal tool, unable to release it from the wires, and feels the steady waves of electricity move from his hands, down his arms, through his chest, and down both legs to the wet floor. His feet are vibrating from the current.

Out of the corner of his eye, he sees the backs of his captain and the two other firefighters. They cannot see him. Their attention is turned elsewhere—they have discovered fire in the ceiling. Fear strikes the firefighter as he realizes his situation: he is being slowly electrocuted.

In a panic, he tries to release his grip and drop the tool, but he cannot. The electric current keeps his wet, gloved fingers wrapped tightly around the metal tool. He tries to shout for help, but he cannot. His jaw is locked shut by the jolts of electricity flowing through his body.

Slowly he starts to shake, with a steady motion, from his fingertips to his toes. The jolts of electricity are beginning to concentrate in his chest. His heart is pounding. The heartbeats are becoming stronger, trying to resist the vibration of the electric current that is taking over his body.

Is this it? Is this the way I am going to die? What a stupid way to go! the firefighter thinks to himself.

He begins to black out. His eyes roll up into his head. Momentarily he loses consciousness. As his body crumples to the floor, his hands are pulled away from the metal tool, and he regains consciousness. "Thank God!" the firefighter cries out as he falls to the wet floor, knocking over several chairs stacked on a table.

Slowly he pulls himself up off the wet floor and staggers outside into the cool autumn afternoon sunlight. He stumbles over to the fire truck and falls down on to the back step. A pump operator runs over to him and asks, "What happened to you? Are you all right?" The firefighter nods his head. "I almost bought it. I almost was electrocuted."

One of the main reasons firefighters are killed and injured at fires is because they work in dangerous, uncontrolled environments—

burning buildings. The air in the burning building is filled with poisonous gases and heat; there is no visibility or light—just thick black smoke—and the entire structure may explode or collapse at any moment.

No other worker in America is subjected to such a dangerous work environment. After a fire has been extinguished and before salvage and overhauling begin, however, the firefighter's environment can be made safer and the work area can be controlled. Fresh air can be pumped into the hot, smoke-filled area by fans and ventilation systems, portable lights can be placed in the area to be overhauled, and a safety survey for signs of collapse can be undertaken before salvage and overhauling begin. Firefighter deaths and injuries during overhauling operations can be prevented.

Perception of Danger

Everything in life is relative—even the perception of danger. To a firefighter who has just risked his life advancing an attack hose line into a blazing home, or who has just searched above a raging fire to rescue a trapped victim, the risks of overhauling seem very minor.

After a fire, the firefighter's perception of danger is terribly distorted. Where the firefighters see a smoldering, charred mattress, moments before they saw deadly flames and felt the radiated heat. Where the firefighters pull down sections of blackened asbestos and plaster ceiling, moments before the same red-hot ceiling was collapsing all around them (Figure 16.1). Where the firefighters see long,

Figure 16.1. *Firefighters should wear self-contained breathing apparatus when pulling ceilings. Hazardous dust particles can be inhaled during overhauling.*

sharp nails and wood lath splinters fly past their eyes and faces, moments before they saw a dangerous suspended ceiling framework which could have collapsed on top of them and trapped them in a small space, in which they could have slowly burned to death or been overcome by smoke inhalation.

After a serious fire, firefighters are exposed to many physical discomforts. The firefighters may be soaking wet, have headaches, and be extremely tired from the exertion of firefighting. The weather may be below freezing or extremely hot, and these climatic discomforts become more noticeable after the exertion of firefighting. There may be emotional discomforts as well. The firefighters may be frustrated or angry by events surrounding the fire or tactics—or, worse, the firefighter working beside him may have been injured.

Discomforts caused by the after effects of firefighting can interfere with decision making and judgment during salvage and overhauling. Time becomes more important than safety to tired firefighters. Exhausted, they trade off their own safety more readily—working in an area in danger of collapse, for example, or not wearing protective clothing—for a quick return to the firehouse for dry clothes, a meal, and some rest. "Get it done and let's get back to quarters" can often be heard during overhauling.

One of the most effective methods of reducing injuries during overhauling is to send the exhausted firefighters who have extinguished the blaze back to quarters and call reserve firefighters to the scene. These fresh firefighters will not risk safety for time; they will not have to remove protective equipment because they are hot and sweaty; they will not underestimate or have a distorted perception of the dangers of the overhauling.

Some General Precautions

After a fire has been extinguished and the arson investigation completed, all the firefighters regroup at the point of fire origin for overhauling duties. At some fires, too many firefighters start overhauling in one small, burned-out room and create a dangerous work environment. It's true that three firefighters opening up walls, ceilings, and floors looking for smoldering, hidden fire will complete the job faster than one firefighter will. Nevertheless, the chance of one of the three firefighters receiving an injury caused by uncoordinated work in a small area is great.

The officer must control and organize the overhauling operation. The hose stream should be temporarily withdrawn from the work

Figure 16.2. *Firefighters should be assigned specific areas to overhaul in order to reduce the danger of overcrowding when working in close quarters.*

area; proper tools should be brought to the area; overhauling should begin at the room of fire origin and work outward. Firefighters should be assigned specific areas to overhaul (Figure 16.2). In a typical residence building, one firefighter can easily overhaul one room. When two firefighters are working in close proximity to each other, the company officer should personally supervise and coordinate the work to prevent injuries. When there are more firefighters available than are necessary to perform the overhauling work safely—without overcrowding—the officer can order that the extra firefighters perform other tasks. The electric and gas utilities to the burned-out area can be shut off. Portable fans and lights can be placed in the area of overhauling, and unnecessary tools and ladders can be replaced on the fire truck.

When a blaze is extinguished, the first overhauling action taken (after safeguarding the furniture) is pulling down sections of ceiling over the fire area to check for hidden flames or smoldering wood beams. During overhauling at a serious fire where several firefighters are using pike poles to pull down sections of large ceilings, there are many dangerous, sharp objects falling around the firefighters' faces, which could cause blindness or a disfiguring facial scar. Firefighters pulling ceilings have been injured by pointed, rusted nails attached to plasterboard, by sharp wood splinters from broken wood lath behind plaster ceilings, by sharp edges of ornamented tin ceilings, by light fixtures swinging down from ceilings, by hanging bx cable, electric conduits, gas tubing, plaster dust, asbestos, and fiberglass insulation. Even the pointed edges of other firefighters' pike poles have struck and injured nearby firefighters. Most firefighters wear helmets equipped with an eyeshield designed to be lowered during this dangerous overhauling stage of a fire. In some instances, foolish though it may be, the eyeshields are not used.

Another cause of injury during the task of pulling a ceiling is the improper use of the pike pole. The firefighter must first check the forward position of the metal hook before he raises it into the smoky upper levels of a room to pull down a ceiling. Next, the firefighter must glance up at the charred ceiling while the officer's light beam is pointed to pick the spot for hook penetration. Finally, the firefighter must look downward and drive the hook point up through the charred plaster ceiling, with several short, sharp, downward pulls. If the pike pole is pulled down too forcefully, the firefighter could lose control of the tool and accidentally strike another firefighter with it.

Removing Smoldering Chairs and Mattresses

Charred upholstered chairs, couches, and cotton mattresses often reignite and flame up after they have been extinguished with hose streams. The cotton padding exterior prevents water from penetrating the interior of the piece of furniture. The combustible wood frame, the horsehair, sisal, or other stuffing used in the interior, together with the air space created by the inner springs, allow a hidden spark to smolder and reignite hours after the outer surface fire has been extinguished. There have been tragic instances in which a fire in a chair or mattress appeared to be extinguished during overhauling and was left in the house. After the firefighters returned to quarters, the stuffed piece of furnishing ignited for a second time, creating a second, larger fire, and taking lives.

Because these stuffed furnishings have often reignited after being overhauled and after firefighters have left the scene, many fire departments have introduced an overhauling policy of removing these cotton-padded stuffed chairs, couches, and mattresses to the street. There they are cut open, pulled apart, soaked by a hose stream on the inside and outside, and, if necessary, broken up. There is nothing as damaging to a fire department's reputation as the rekindle of a fire after firefighters have left the scene.

The sudden reignition of a stuffed chair or mattress can injure firefighters. There have been many instances in which a smoldering mattress being carried down a stairway to the street suddenly reignited, flamed up, became stuck in a stairway or doorway, and burned a firefighter.

Any cushioned chair or mattress which is smoldering in a hot, damp, smoky apartment can suddenly burst into flame when carried down a stair or out through a hallway where cool, fresh air flows around it. Before a smoldering mattress or cushioned chair is carried or dragged outside, it should be cut open and the interior soaked with water. If this work cannot be done, a portable extinguisher or hose line should be ready to quench any reignition of the furniture as it is carried outside.

For these reasons, smoldering mattresses or stuffed chairs should never be taken down to the street in an elevator. There have been cases in which building superintendents have attempted to remove smoldering furnishings from upper floors of high-rise buildings via inside elevators. They were trapped in the elevator car by the reignited mattress and burned to death before the flaming car reached the lobby.

Another dangerous task carried out during overhauling is throwing a smoldering mattress or stuffed chair out of a window. Dropping rubble from a window down into the street or yard without communicating that intent to firefighters below could cause death or serious injury. The falling piece of furniture may strike a firefighter who is just leaving the fire building (Figure 16.3). Even fire rubble thrown out a rear or side window could injure a firefighter about to climb a fire escape or raise a ladder.

One evening, firefighters entered a multiple dwelling for a smoky fire on the fourth floor. The fire in the living room was quickly extinguished, the room vented, and a victim dragged out of the apartment. The smoldering chair was quickly carried over to a rear window, maneuvered out on to a rear fire escape, and tossed over the railing into the dark backyard. Down in the backyard, another firefighter from a second-arriving ladder company was moving to lower the fire escape "drop ladder" and climb the rear fire escape. He was struck by the falling chair. He suffered brain damage and a permanent loss of balance when walking.

If it is absolutely necessary to throw a piece of rubble out of an

Figure 16.3. *Pieces of burned furniture thrown from windows during overhauling can seriously injure firefighters working around the perimeter of a building.*

upper-floor window, the firefighter must first contact another firefighter at street level. Notify him of the intended action. Request that the area be cleared and that no one be allowed to pass beneath the planned landing place. Only after receiving the "all clear" from below should the smoldering rubble be thrown out the window. Without some contact with a firefighter at street level, no object should be thrown out of the window. Shouting "Watch out below!" is not sufficient.

Carbon Monoxide in Below-Grade Areas

After a fire has been extinguished and before overhauling begins, firefighters are often sent to the cellar areas of the fire building to shut down affected utility services or what has now become unnecessary fire protection equipment serving the burned-out store or apartment. Gas meters are shut off if the pipes are broken; electric meters are shut off where water has caused electrical wires to short circuit; sprinklers are temporarily shut off in the cellar in order to change a fused sprinkler head. Firefighters sent to cellars to perform these duties without self-contained breathing equipment have died and their predicaments have caused the deaths of other firefighters coming to their rescue.

Consider the possibility: A firefighter who descends into the cellar without a mask falls unconscious, overcome by accumulations of carbon monoxide from the fire that's just being extinguished (Figure 16.4). After realizing that he's missing, two or three of his buddies rush to the cellar—also without masks—and are quickly overcome.

Finally, a firefighter enters the cellar with self-contained breathing equipment strapped on and operating. An emergency call is radioed to the firefighters above and the unconscious firefighters are dragged up out of the cellar. Their bodies are examined. The degree of brain damage and the number of dead firefighters will depend upon the sequence in which each maskless firefighter was dragged out—and the time it took for the first firefighter with his SCBA in place to enter the cellar.

Firefighters descending cellar stairs during overhauling must consider the possibility of carbon monoxide and smoke accumulation in the cellar at all times, but particularly so when a fire of long dura-

Figure 16.4. *Self-contained breathing apparatus should be worn when firefighters enter a cellar, after a fire has been extinguished, to shut off utilities.*

Figure 16.5. *Metal ladders or tools contacting electric wires are the cause of firefighter electrocutions.*

tion has just been extinguished, when the fire location is a store on the first floor, directly above the cellar, and when the cellar is completely below grade.

Electricity

Research into electrical danger on the fireground is misdirected and misleading; it concentrates mainly on the conduction of electricity through hose streams. Nevertheless, most firefighters who are electrocuted are not directing hose streams—they are holding a metal tool or a metallic piece of equipment that comes in contact with live electrical equipment.

Firefighters carrying metal tools and climbing metal ladders near electrical equipment are in danger of electrocution. When a metal tool or ladder accidentally comes in contact with live electric power, the firefighter's body completes an electrical circuit (Figure 16.5). Current is relayed to ground through the firefighter's body still more readily when his clothes are wet and when he's standing on a wet surface.

During overhauling, both firefighter and floor are wet. The firefighter is using metal tools. Firefighters should treat all electrical equipment as live and should avoid coming in contact with it during overhaul. If sparks or shocks from electrical equipment are received, the officer should be notified of the danger and the electric supply shut off. Before commencing to shut off any electric supply, a firefighter should be equipped with non-conductive gloves, his eyeshields should be in place, and he should be standing on a dry, non-conductive surface.

Another electrical danger to firefighters during overhauling is arcing. Arcing is the situation in which a large electric spark jumps between two closely spaced, conductive objects when electric current is interrupted. One of the conductive objects could well be a firefighter. A spark might jump from an electrical supply panel into a firefighter when he is shutting off the electrical supply and standing on a wet, conductive floor.

Arcing often occurs when a switch is opened or when fuses are pulled to interrupt or to shut off electric power. The arcing of electricity could severely burn a firefighter.

Falls

According to National Fire Protection Association statistics, falls are the second leading cause of firefighter deaths and injuries on the fireground. If you survive fighting a fire and begin to overhaul the burned-out area, chances are great that you may still die or be injured by a fall outside or inside the fire-damaged structure (Figure 16.6).

The dangers of falls during overhauling operations are present both outside and inside a burned-out structure. Outside the structure, ice is the major cause of a firefighter losing his balance and falling. On a cold night, water from hose streams freezes. A firefighter who stretches a hose line up a couple of dry front steps to attack a fire or who climbs up on a peaked roof to cut a vent opening may experience no difficulty with footing, but he may walk back down those same steps or take the same path down the sloping roof only to slip and fall on water from the hose streams that has turned to ice.

Inside a burned-out building, ice is not usually the problem. The problem is one of perception: all the walls are black from the soot, smoke, and char of the recently extinguished fire; everything looks the same. Normal warning signs, visible before the fire, are now destroyed (Figure 16.7). Signs on windows warning of shaftways are burned or missing. Hinged elevator doors look the same as apartment doors. A black hole in a floor looks the same as a charred black rug. Window frames and doors which might keep a person who trips from falling out of the building are missing, burned away, or removed. Everything is dark, utilities have been shut off, and even the flames from the extinguished fire no longer reveal a missing stair.

Here are some examples of how firefighters have been injured by falls in and around smoke-filled buildings during overhauling.

- A firefighter stands on a box spring and mattress. The room suddenly bursts into flame, and he tries to take cover below the windowsill. Movement of the box spring and mattress cause him to fall out the window.
- A firefighter about to throw a heavy, wet mattress out a window catches a piece of the inner spring on his turnout coat clip and is pulled out the window by the falling mattress.
- A fire company discovers flames traveling around a partition into an adjoining apartment. A firefighter is ordered to go to the adjacent apartment and examine it above the ceiling for fire spread. As the door from the dark, smoky, burned-out hallway next to the fire apartment is opened and entered, the firefighter falls into an elevator shaft.
- At night, a firefighter opens the door on a roof structure which appears to lead to an interior stair. He enters; it is an elevator shaft. He falls 11 stories to his death.

Figure 16.6. ABOVE: *The danger of falls during overhauling operations are present both inside and outside the burned-out building.*

Figure 16.7. *During overhauling after a serious fire, safety devices and warning signs are burned away or charred by smoke. Set up portable lights during overhauling.*

- A firefighter opens the door to a roof structure of a vacant building which leads to the interior stairs. He steps inside and falls five stories. All the landings, treads, and risers of the stair to the roof have been removed by scavengers.
- A firefighter steps from an aerial ladder into a burned-out window. There is no floor—it is a shaftway. The "Danger—Hoistway" sign was burned off the window during the fire.

The firefighter's best protection against suffering a fall on the fireground is a flashlight. A flashlight affixed to a turnout coat or helmet can free the firefighter's hands to carry tools and reveal dangers.

The firefighter must understand the danger of the "moth and the flame" syndrome: he becomes vulnerable to injury when he is concentrating on the flames and not on the surrounding dangers. In the same way, a firefighter given an urgent assignment during overhauling often becomes the victim of a danger that would have been easily seen and avoided during non-emergency conditions.

Floor Collapse

When a serious fire involves several floors of a house and outside streams are required to extinguish the fire, firefighters must sometimes be sent inside to complete a secondary search for victims and to overhaul the gutted building.

The most dangerous floor collapse area inside such a burned-out structure is the bathroom. When a firefighter enters this room to search or to overhaul, the floor joists may suddenly fail and cause the collapse of the entire bathroom floor. The firefighter, along with heavy cast-iron sinks, bathtubs, porcelain toilets, and heavy tile floors, will crash into the basement or the floor below.

There are several reasons why the bathroom floor is more susceptible to collapse than other floor areas.

- The bathroom fixtures create a heavy dead load. Cast-iron tubs and sinks and porcelain toilets can weigh up to 1,000 pounds. This weight is concentrated in one small area.
- In some older buildings, the thick tile and sand-bed floor insulation required that the floor beams be reduced in size (Figure 16.8).
- Over the years, the moisture from sweating or leaking water

Bathroom Floor Construction of Older Structures

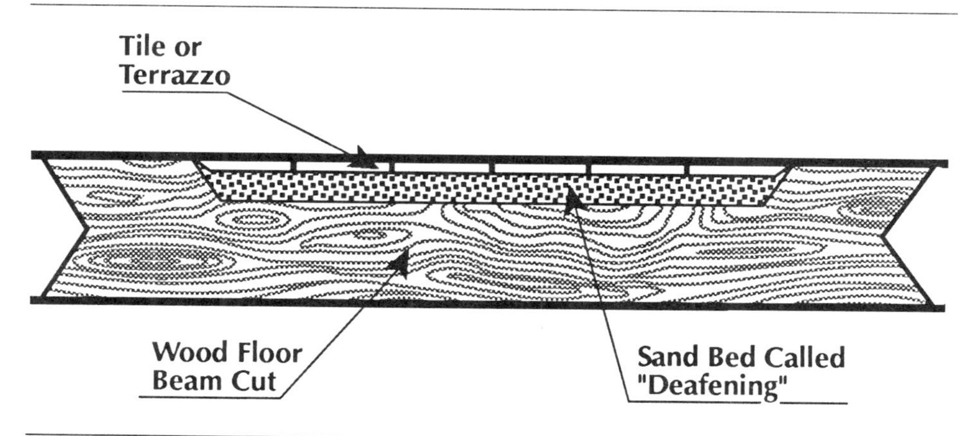

Figure 16.8. *Bathroom floor insulation, called "deafening," may cause reduction of the size of the floor beams in addition to adding to the dead load.*

pipes can cause the wooden bathroom floor joists and floor beams to rot or weaken.

- The bathroom floor joists are more likely to be destroyed by fire because the bathroom often contains the most poke-through holes and concealed avenues of fire spread. Fire burning upward from one floor to another will seek out the path of least resistance: the bathroom floor. As the flames spread up through bathroom poke-through holes, they weaken the floor joists.

During overhauling operations in a seriously damaged bathroom, firefighters should attempt to open up a fire-weakened ceiling or wall using the reach of a pike pole and standing outside the bathroom. Firefighters using a hose stream to wet down a burned-out bathroom should also stand outside the room and use the reach of the hose stream to extinguish any small pockets of fire.

Gas Explosions

When a fire company extinguishes a smoky room and contents fire in a kitchen or basement, the hose stream may accidentally extinguish a gas fire from a melted or broken gas pipe, tube, or meter. The unburned gas leaking into the charred room may be suddenly reignited by a smoldering ember and explode violently.

The time between extinguishing a house fire and before overhauling begins is a dangerous stage of such a fire. Gas explosions often occur at this time. Gas meters melt, copper tubing softens and separates, and pipe joints come apart when ceilings collapse.

The possible presence of leaking household gas must be determined after a fire has been extinguished and before overhauling begins. Both natural gas and bottled gas are required to be odorized so that people can detect them at gas concentrations in air not exceeding one-fifth of the lower limits of flammability. This odorization is not, however, effective in smoke-filled rooms for firefighters wearing SCBA.

The most effective action a firefighter can take immediately after a fire has been extinguished is to prevent an explosion by venting the fire area (Figure 16.9). Next, any gas appliances should be checked for leaks before overhauling. Some fire officers carry putty or clay to plug up broken or leaking gas pipes quickly; a rag or soap can be used temporarily to plug up escaping flammable gas, while other firefighters are simultaneously venting windows. It is erroneous to assume that a flammable gas-air mixture needs to fill a room or building completely before an explosion can occur. Most combustion explosions inside buildings occur with less than 25 percent of the enclosure filled by the flammable gas-air mixture.

If an explosion of reignited gas does occur, the gas burning at the appliance should be allowed to burn until the flow of gas to the broken pipe has been shut off. The room and contents fire should be extinguished, and the exposures protected.

Lacerations and Cuts

The danger of falling glass is always present on the fireground. Severe facial, hand, shoulder, and back lacerations occur when broken pieces of window glass fall on firefighters from the upper stories of a burning building. The most dangerous area of the fireground for such injuries is the perimeter of the burning building. Firefighters are most often cut by falling glass when walking into or out of the building, when operating on a ladder placed at a window, or when operating on a fire escape.

After a fire has been extinguished and overhauling has begun, a survey of the broken windows will reveal large, jagged pieces of broken glass resting on the outside windowsills and hanging from or still remaining in the window frames. The upper portion of a store's large, glass display window may still be in its frame after the lower portion has been broken out during venting. These large pieces of

Figure 16.9. RIGHT: *After a fire and before overhauling, vent the building to prevent the build-up of combustible gases.*

Figure 16.10. *Broken pieces of glass on a porch roof can create a slippery surface.*

glass present a deadly guillotine hazard to anyone who walks through the broken window. They should be removed immediately, broken by a firefighter using a pike pole from a safe distance.

All windows of a burned-out building should be stripped of any remaining glass shards during overhauling—but not until warning has been given to firefighters below. Large pieces of broken glass on the outside windowsills of upper floors should be removed carefully by a gloved hand. Jagged pieces of glass remaining in the window frame should be removed by a pike pole, axe, or halligan tool. The glass pieces should be tapped back inside the broken window, not outside.

Do not linger around the perimeter of a burned-out building; go inside or move far enough away from the front of the building not to be struck by falling glass.

Broken pieces of glass window on a concrete sidewalk, metal fire escape, or slate porch roof will create a slippery surface (Figure 16.10). Firefighters stepping out of a second-floor window at night on to a porch roof covered with layers of broken glass may slip and fall off the roof. Portable ladders should never be placed on top of broken layers of glass on a concrete sidewalk.

Heat Exhaustion and Overhauling

While performing rescue and hose line attack during the fire, firefighters are exposed to three types of stress: physical stress from hose stretching, forcible entry, and raising ladders; emotional stress from the fear of death or serious injury caused by flashover, explosions, and collapse; and heat stress caused by exposure to the high temperatures of the flames, heated smoke and gases, ambient temperature, and the insulating effects of their protective clothing.

Once a fire has been extinguished and the dangers are reduced, the physical and emotional stresses are no longer great; however, the heat stress continues and, for some firefighters, actually worsens during overhauling. After the tremendous effort expended during the fire and rescue operation, the firefighters regroup inside a hot, steamy, burned-out structure to start strenuous overhauling, looking for fire extension, extinguishing spot fires, and removing burned rubble from the fire area (Figure 16.11).

It's then that heat stress begins to affect firefighters. They may suffer heat cramps, heat stroke, or heat exhaustion caused by elevated body temperatures. Statistics reveal that firefighters most often suffer heat exhaustion. These conditions are recorded as injuries caused by

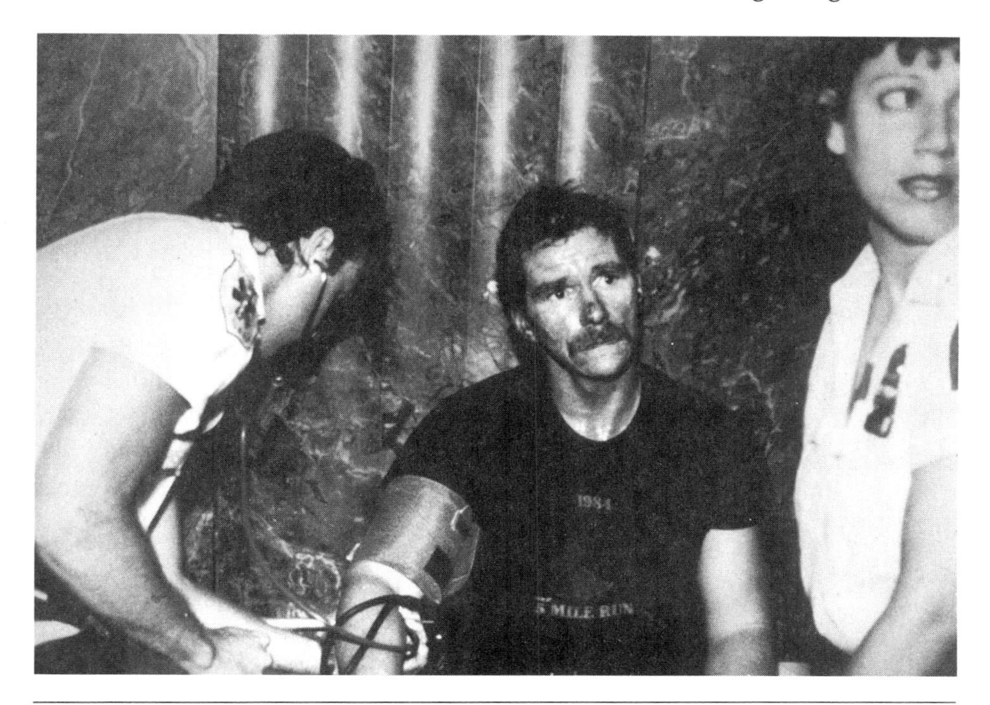

Figure 16.11. *The effects of heat and physical exertion during overhauling can cause death and injury.*

overexertion on the fireground—the fourth leading cause of firefighter injury.

Rookie firefighters are particularly susceptible to this type of injury. Trying to prove themselves as firefighters, they overexert themselves both during and after a fire. The symptoms of heat exhaustion catch up with them during the overhauling stage. Veteran firefighters should know how to pace themselves and save that extra bit of energy for the overhauling operation.

When a firefighter begins to experience heat exhaustion, he sweats more than is normal, may feel a tingling in his arms or legs, turns an ashen or gray color, has difficulty in breathing, feels nauseated, and possibly vomits. Any of these symptoms are signals to slow down. Notify your officer that you need a break. If possible, go back to the apparatus and get a drink of water to replenish the water you lost from perspiration, and take off your helmet, protective hood, and turnout coat. Splash some water over your head and into your face, and sit down on the back step of the fire truck. Don't try to fool yourself, your supervisor, or your department—take a break from firefighting. It's better to take a short rest than to drop unconscious inside the burned-out fire structure.

Even if you are able to go through the motions of overhauling while feeling exhausted, you may cause an injury to a fellow firefighter. It's a well-known fact that exhausted and overexerted firefighters make poor judgment decisions on the fireground. Many fireground injuries during the overhauling stage of firefighting are caused by firefighters, officers, and chiefs whose judgment has been impaired by exhaustion.

Be aware of the condition of firefighters around you. If they show signs of heat exhaustion, slow down and inform your officer.

An officer and a chief should be able to recognize different levels of exhaustion on the fireground and take appropriate action. Rotation of firefighters, short rest breaks, and calling fresh firefighters to the scene to handle overhauling can often reduce injuries and is good firefighting strategy.

Lessons to Be Learned

Before overhauling operations begin, there are several safety precautions a fire officer can order to reduce the dangers:

1. Conduct a collapse danger survey and rope off all danger areas.
2. Set up portable fans to remove toxic smoke and gases.
3. Set up portable lights with their own power supply.
4. Shut off gas and electric utilities.
5. Determine the fitness of firefighters, and select those most able to perform overhauling. Rest and rotate the others.
6. Assign firefighters specific areas that are large enough to provide safe work spaces.
7. Supervise and coordinate overhauling in close quarters where two or more firefighters must work together.
8. Require that SCBA facepieces and eyeshields be worn when pulling ceilings.

17

Risk Analysis Dangers

"We've got a job," the officer says to the driver. Reaching back to slide open a window between the cab enclosure and the jump seats where two firefighters are riding, the captain repeats, "We've got a job; mask up."

Pulling up next to a burning farmhouse, the pump operator says, "Captain, we're in luck. There's a hydrant across the road from the house. Let's make a steamer connection."

After sizing up the fire building from the cab, the captain radios a preliminary status report. "Engine One to Communications Center."

"Communications Center to Engine One, go ahead with your message."

"Engine One, we have arrived at number two Maplewood, 200 yards north of Route One. We have heavy fire and smoke showing from the first floor of a two-and-a-half-story, 20- by 40-foot wood-frame house. Exposure number two is a one-story frame garage. Send a backup mutual-aid engine company from the next county."

"Communications Center to Engine One, ten-four. The estimated time of arrival for the second engine is 20 minutes."

"Engine One, ten-four."

"Stretch!" the captain shouts to the two firefighters as he leaves the cab and runs toward the burning house. Flames and heat have broken the two windows and the glass panels in the wood door that leads out to a large front porch. Flames fan out beneath the porch ceiling and smoke billows up around the outer edges of the porch roof, obscuring the top story and the attic of the house. Light bulbs on a porch ceiling fixture are exploding from the heat of the flames. As the officer is about to climb the two steps leading to the porch deck, he hears a voice from above. "Up here, I'm up here on top of the porch. Help!"

Stepping back from the house and shining a handlight upward through the smoke, the captain can barely see an elderly man standing on the porch roof, holding a handkerchief to his face, waving his arm. "Get a ladder! There's a guy on the porch roof," the officer calls back to the firefighters stretching the pre-connect attack line.

Turning back to the trapped man, the captain yells, "Don't jump. Stay where you are. We'll get you."

The three firefighters at the scene work quickly and independently. One firefighter removes a 20-foot ladder from the side of the pumper; the other firefighter, holding the nozzle, kicks and pulls at the uncharged hose line to prevent kinks that would reduce the nozzle pressure. The pump operator, bending over the hydrant, connects the pumper hose to a large outlet.

Arriving at the porch with the ladder, the firefighter, in one quick, uninterrupted motion, places the butt end of the ground ladder against the porch foundation, raises it vertically, hand over hand by the rungs, pulls the base out from the porch to set the proper climbing angle, and starts up the ladder through thick, black smoke. The officer butts the ladder. Once on the roof with the elderly man, the firefighter places one arm around the man's shoulders and leads him to the ladder at the edge of the roof.

"I can't see well at night! Don't let me fall," the man cries to the firefighter.

"Don't worry; you'll be okay. Just follow me." Flames suddenly appear above the porch at the other end of the roof. At the edge of the porch roof, the firefighter assists the man on to the ladder. "Here we are. Place your right hand on the right beam." The firefighter guides the man's hand in the darkness and smoke. "Now lean forward and place your other hand on the left beam." The firefighter steadies the ladder by holding the top rung. "Good. Now step around and place your left foot on the rung. Shift your body weight to the leg on the rung. Great. Now swing your right leg around and place it on the same rung. That's it; you got it!"

By this time, the captain has climbed the ladder and is positioned behind the man. Using his body and arms, the captain encircles the man to prevent him from falling backward down the ladder.

"Okay. I've got you. Come on. Let's start down the ladder," the captain says. "Move your left foot down to the rung just below. That's it. Now the other foot. Great! Let's keep going."

"I can't! The smoke is choking me!" the man shouts as he starts to climb back up the ladder. Pressing his body against the man, the captain forces him against the ladder rungs to prevent him from climbing back up. Flames start to creep down the edge of the porch roof, closer to the ladder. The captain talks into the man's ear as smoke engulfs them both.

"Hold your breath, turn your head away from the smoke, and let's go before it gets worse."

The man starts back down the ladder, the officer guiding him on the way. Once safely on the ground, the captain asks, "Is anyone else inside?"

The man replies, "Yes, my dog. Please save my dog. His name is Rusty!"

"Okay, okay; anyone else?"

"No," the man replies.

Suddenly, the smoke rising up around the ladder flashes into flame. The firefighter is trapped on the porch roof, encircled by a ring of fire. Unable to climb down the ladder, the firefighter quickly moves across the roof, picks out a spot where the flames are less intense, and jumps off. The captain runs over to help the firefighter up, "Are you all right?" he asks. The firefighter, out of breath, nods his head yes.

The officer and the firefighter regroup around the hose line, now charged with water. The nozzle is held by a firefighter who is masked and ready to attack the fire. "Hit the fire from here," the officer tells the firefighter standing on the step of the porch deck. The firefighter opens the nozzle and sweeps the under side of the porch with the powerful hose stream several times. The flames beneath the porch turn back into smoke.

"Now hit the fire in the windows and doorways," directs the officer. The stream crashes through the broken windows and darkens down the flames inside. As the stream strikes the front door, it swings inward violently and crashes against the inside wall. The flames are driven back, and the officer shines his light on the porch supports, the columns and ledge beam where the roof meets the house. "Okay," affirms the captain, "let's move inside the doorway."

The firefighter directing the nozzle, followed by the captain, moves up on to the porch, crosses it, enters, and crouches down several feet inside the doorway, directing the stream into the rooms off the hallway. Another firefighter pulls up the excess hose for the nozzle advance and then also moves up into the doorway.

Outside the burning building, the pump operator checks the gauges, verifies that inflow and outflow pressures are stabilized, grabs a six-foot pike pole from the side of the pumper, and runs around to the rear of the house to vent windows for the advancing hose line.

At the back of the house are an outside cellar entrance door and three windows. The firefighter pulls open the cellar entrance door to check if the fire originated below grade. As the door opens, a soaking-wet dog darts out of the darkness with his tail between his legs. There is no indication of heat or smoke coming from the cellar.

The nearest of the three windows is discolored by heat and about to crack, with smoke pushing out around the window frame. The pump operator passes this window and breaks the window farthest downwind first. With the top of the pike pole, he taps the six topmost glass panes first, then breaks the large glass bottom pane. Smoke pulls out of the vented window. Backing up, the pump operator decides not to vent the middle window because a propane gas cylinder is visible near the opening. He stands next to the first blackened, heat-discolored window and is raising the hook to break it when suddenly the entire window explodes outward, water spray and flying glass narrowly missing the firefighter's face. The 30-degree fog stream shoots out the broken window along with volumes of smoke and heat.

Venting completed, the pump operator runs back to the fire truck, replaces the hook, and checks the gauges again. Fire appears to be extinguished on the first floor, but flames are now shooting out two second-floor windows, lapping up near the attic window and the peak of the gable roof. The entire front of the house and the path leading to the porch are aglow, lit by the bright orange flames blowing out the windows and the revolving red light from the pumper.

The pump operator climbs on top of the pumper behind the cab, grabs the long barrel of the fixed portable deluge nozzle, and pushes and pulls it around, pointing it toward the flames and the second floor. The firefighter climbs back down the side of the pumper, pulling down with him the short length of hose pre-connected to the deck pipe. (The hose is connected to an outlet on the pump.) As he watches the flames spread up to the attic window and eaves, the pump operator thinks to himself, "If the company can't make the second floor with the interior attack line, the deck gun will be ready. I'll be all set if the captain gives the order to back out and fight it from outside."

The sounds of the pumping engine and the inside attack stream crashing against the hollow plaster and lath walls drift across the open field from the burning farmhouse. The pump operator, sizing up the front of the house fire, thinks, "I could knock those flames down in one or two seconds with that master stream on top of the fire truck. I could also set up a water curtain by directing the deluge nozzle stream against the spandrel wall—the wall space between the top of the flaming second-floor windows and the bottom of the attic windows." But, he also thinks, "I could miss the spandrel wall and drive the stream into the second-floor windows and burn my guys inside. The control of an outside master stream is not very precise. The stream trajectory is not usually accurate, and the water pressure is often erratic when the stream starts up. Besides, the captain has given strict orders never to direct the deck gun into the building when the company is operating inside."

Suddenly, the flames in one of the windows are snuffed out. White smoke blows out the opening. Then the flames in the other window quickly disappear, replaced by billowing clouds of white smoke. The front of the house returns to darkness; only the red lights of the apparatus flash against the front of the building.

The captain leans out one of the smoke-filled windows, pulls off his face mask, and looks up at the attic window. Small flames flicker near the attic window frame and around the roof peak. A firefighter climbs out of a smoke-filled window and steps out on to the porch roof. The nozzle and hose are passed out the window to the firefighter, who sprays the small flames near the attic window and the roof peak (Figure 17.1). The beam from the officer's handlight can be seen inside the attic window through the soot-covered glass. The firefighter on the porch roof turns to the pump operator, directing the apparatus spotlight beam up to the second-floor window, and gives a "thumbs-up" sign. The elderly man sits on the back step of the pumper holding his wet dog, and the siren of the second responding engine company can be heard in the distance.

Risk-Taking Decisions

Most firefighters killed in the line of duty die performing routine firefighting tactics, not daring rescues; the routine tasks of firefighting are dangerous. Crawling through a smoke-filled house searching for the fire origin is dangerous (Figure 17.2); advancing the initial attack hose line to the fire is dangerous; climbing a ladder to a peaked roof to vent an attic fire is dangerous; even pulling a plaster lath or tin ceiling in a burned-out room is dangerous.

Each year, routine firefighting tasks kill approximately 130 firefighters and injure 100,000 others. When a firefighter dies or is seriously injured, the chief and company officers of that department often re-evaluate the particular firefighting tactic performed. The re-evaluation can occur at an official board of inquiry established by the chief

Figure 17.1. LEFT: *Porch roofs can collapse because of fire destruction or overloading by firefighters.*

Figure 17.2. *The first officer entering a burning building to conduct an interior size-up is performing an extremely dangerous firefighting tactic.*

or it can simply be a moment of silent anguish, asking, "What happened at that fire to cause that firefighter's death?"

Asking ourselves that question many times has led to risk analysis of firefighting tactics. Fireground risk analysis is the method we use to identify, evaluate, and control firefighting dangers. An important part of risk analysis is establishing the three priorities of firefighting risk-taking (Figure 17.3).

Establishing Priorities

A firefighter should expect or accept a degree of risk during a routine firefighting operation. The degree of risk varies, depending on

Priorities of Risk Taking

1 **Protection of Life**
 (a) **Civilian**
 (b) **Firefighter**
2 **Fire Containment**
3 **Property Protection**

Figure 17.3. *All firefighting decisions can be made using these firefighting guidelines.*

whether the firefighting operation is intended to save a life or to contain a fire. The same degree of risk taken by a firefighter to save a person trapped by flames or smoke need not be taken in a routine fire containment operation.

How far should I search into this smoke- and heat-filled store? When should I back a line out and set up an outside attack? When should I get off this fire-weakened roof? These questions can be answered—and every life-and-death decision made on the fireground can be made according to the three firefighting priorities of risk-taking.

The protection of life is the highest goal of the fire service. During a fire or emergency, firefighters often endanger their own lives to save the life of another. When a life is clearly threatened, there is no risk too great. At most fires, however, lives are not clearly endangered. At most fires, then, the priority of firefighting is the protection of the firefighters' lives.

Firefighters risk their lives to a great degree by exposing themselves to flame and smoke during a routine firefighting tactic. When no other person's life is in danger, the life of the firefighter has a higher priority than fire containment.

For example, in the account that introduced this chapter, when the engine company arrived at the farmhouse, there were flames shooting out the front windows and door, and an elderly man was trapped on the porch roof. The life of the man was first protected by stretching a hose line and raising the ground ladder. If the firefighter climbing up on the porch roof had been killed or seriously injured by a porch collapse, the risk taken to save the man would have been justified. The goal of every action in the fire service is to protect lives from fire.

After the man had been removed from the roof, the officer directed the firefighter operating the nozzle to sweep the fire with the hose stream before entering beneath the porch roof and checked the porch supports for stability before advancing the hose line—all actions taken in accordance with the priorities of firefighting risk-taking.

After the hose line was advanced into the burning house and the flames were blowing out of the second-floor front windows, the pump operator positioned the deck gun toward the gun but did not use the master stream. The firefighter's decision not to use the outside stream when his fellow firefighters were operating a hose stream inside the burning structure was, again, in accordance with firefighting risk-taking priorities. The safety of the firefighters was considered before fire containment. The master stream could easily have extinguished the flames in the front rooms of the second floor, but it could also have severely burned the firefighters by driving heat and flames into their faces.

Lessons to Be Learned

If the protection of the public is the first priority of risk-taking, and fire containment is the second priority of risk-taking, why should firefighters risk entering a burning building? Why not let it burn and fight the fire from outside the building?

When firefighters enter a burning building and quickly extinguish a fire with the initial attack hose line (Figure 17.4), this action is the most effective method of accomplishing the first and second priorities of risk-taking. Extinguishing the fire or confining it to the area of origin often protects people trapped above, below, or adjacent to the fire. It also exposes fewer firefighters to the dangers of firefighting (Figure 17.5).

Fewer firefighters are endangered during an interior attack on a fire when it is still of manageable size. If the fire is allowed to grow and then fought from outside, more firefighters could be killed or injured by the dangers of exterior firefighting.

Figure 17.4. RIGHT: *Most fires in America are extinguished by the initial attack hose line.*

Figure 17.5. Risk of firefighter injury compared to the benefits of protecting property.

Risk/Benefit Analysis

1 The average replacement cost of a burning building:

2 The yearly cost of medical pension or compensation for a firefighter who is permanently disabled:

3 Total cost of disability payments for 30 years*:

*35 years – average age at time of injury, living to 65 years @ $15,000 = $450,000

4 Compare 1 to 3

18

Interior Firefighting Dangers

AROUND MIDNIGHT, a ladder company arrives at a new, two-story apartment house near the outskirts of town. Fire has been reported in a second-floor apartment. As the captain and two firefighters run upstairs to the second-floor landing, they see a young woman, dressed in a bathrobe, and a building superintendent holding a handkerchief to his face. The woman is crying.

She runs toward the firefighters, pleading, "Oh, please help me! My daughter is still inside the apartment! The living room is on fire! I fell asleep watching television, and when I awoke, the room was on fire!" The building superintendent adds: "I tried to go inside the apartment, but the heat and smoke were too much."

As they all move down the hall, thick, black smoke is pushing out around the top of a partly opened door. The officer crouches down, pushes open the door, and looks inside. Black smoke fills the entire apartment except for three feet at the floor level. A red glow is coming from behind the door to the left within the apartment. The captain raises his hand into the smoke and can feel extreme heat through his glove. The apartment is about to flash over.

"Where is your daughter's bedroom located?" asks the officer. "Straight back! All the way in the rear of the apartment. She is only six years old! Please!" cries the mother. "Mask up," orders the captain. One of the firefighters, holding a portable 2½-gallon extinguisher and a six-foot pike pole, puts the pole down and pulls the lock pin on the extinguisher, readying it for use. The other firefighter, carrying an axe and a halligan tool, chocks open the self-closing door, using the axe head as a wedge between the bottom of the door and the floor. The officer and the two firefighters crawl into the burning apartment.

The rooms are arranged in an open design, with no doors between the living room, kitchen, and hallway. If the apartment flashes over, no one will get out in time. From the center of the apartment, the officer and firefighters can see a large burning sofa in the living room. Flames rise up into a black cloud of smoke. Small flaming droplets fall from the arm rests of the sofa on to the floor. The rug around the sofa is also burning, with a ring of fire moving outward on the fabric.

The captain looks to the rear of the apartment, where the child's room is located, and then at the firefighter with the halligan tool. He taps the firefighter on the shoulder and points toward the closed bedroom door. They split up; the captain and the firefighter with the extinguisher move toward the fire. The other firefighter crawls toward the rear.

At the bedroom door, the firefighter reaches up into the smoke for the doorknob, turns it, and pushes the door open. The light, hollow-core wood door flies open and bangs against the wall inside the dark bedroom. Guided by the bedroom wall, the firefighter searches for the child's bed. He feels the soft mattress and pats the bed with his hand, feeling a small body. The startled child awakens and bolts out of the bed into the arms of the firefighter. The girl's arms go around the firefighter's neck, and the firefighter's arms encircle her waist. Crawling back to the door opening, the firefighter crashes into a blank wall, the child falls to the floor. "The door! The door has got to be here! I know it!" says the firefighter to himself. Scooping up the little girl again, he slides a hand across the wall surface in the dark. His hand bangs against the doorknob. "The door must have closed again after I entered the room," thinks the firefighter. He turns the knob quickly and pulls the door open. The firefighter dashes past the fire with the little girl.

As they exit from the burning apartment, the captain and the other firefighter are still inside, struggling to extinguish the burning couch before it causes the entire apartment to explode into flames. Dragging the extinguisher around the flaming sofa, the firefighter stumbles, crawls, falls, and kneels, attempting to stay below the layer of superheated fire gases while spraying the flames with the small water nozzle. In the blinding smoke and intense heat from the flames, the firefighter points the nozzle at the couch and quickly tries to cover the entire burning surface with water before the extinguisher runs dry. With the nozzle inches away from the flaming upholstery, the firefighter places a gloved finger in front of the nozzle. (This technique breaks the stream into spray and extinguishes fire more efficiently.) The firefighter squeezes the handle of the extinguisher with one hand and directs the nozzle with the other.

Suddenly, the window drapes behind the sofa catch fire. Flames rise up to the ceiling and fan out over the firefighter's head. The captain rips the drapes off the window, throws them on the floor and stomps on them several times, smothering the flames.

"That's it, captain. The can is empty," says the firefighter through his SCBA facepiece. "Go get a pot of water from the kitchen. I think we got it," replies the captain as he reaches behind the smoldering sofa, removes a window shade, feels for the lock, opens it, and pulls down the top half of the double-hung window. Smoke and heat in the apartment rise rapidly. The superheated gases are sucked out the window as fresh air blows into the apartment from the hallway.

"Captain, where are you?" questions a voice in the smoke. "The little girl is okay. She's with her mother," continues the voice. Replies the officer, "Grab that pike pole on the floor and put a hole in the ceiling over the couch. I want to make sure the fire did not get into the ceiling." The firefighter returns with a pot of water and pours it over the smoldering sofa. The ceiling is opened and, as the captain examines the charred material, the portable radio blares: "Engine 8 to Ladder 6. What do you guys have up there? Do you want a hose line?" The captain answers: "No, hold the line; we got it. It was just a couch."

After a firefighter is killed or injured, people outside the fire service often question why firefighters must go inside burning buildings to extinguish fires. There are several reasons why firefighters enter burning buildings, the most important of which is saving lives.

Reasons for Interior Firefighting

During the first few minutes of a hose line attack inside a smoke-filled building, unconscious men, women, and children are tripped over, stumbled upon, bumped into, crawled over, and rescued by firefighters (Figure 18.1). These fire victims would die if firefighters fought fires—more safely—from outside burning buildings. Many lives are saved accidentally during interior firefighting. Most fire victims are discovered by chance, when firefighters are performing routine interior firefighting tactics, such as searching for the fire origin, advancing a hose line, and venting windows from inside a building. Each year 6,000 to 8,000 people die in fires. This statistic, from the

Figure 18.1. *If firefighters did not aggressively attack structural fire, using interior attack strategies, the annual life loss from fire, 6,000 to 8,000 persons, would double.*

Figure 18.2. RIGHT: *Interior firefighting tactics save lives and protect property.*

National Fire Protection Association (NFPA), could double if firefighters did not enter burning buildings.

Firefighters also enter burning buildings to fight fires because of the moral obligation to protect property from destruction (Figure 18.2). Most firefighters—volunteer and career—take an oath to protect life and property when they join a fire department. If we examine the sworn oath, the actual words of the vow may simply state: "obeying laws of the local government," If, however, we read further, the law of the local government mandates that the fire department protect the lives and property of people in the community. So, in exchange for the privileges, compensation, and prestige of calling ourselves firefighters, we swear to risk our lives to protect the houses where people live, and the factories and offices where people work, in addition to saving lives.

Over eight billion dollars worth of property were destroyed by fire in the United States in 1989, according to the NFPA. This monetary loss could increase tenfold if firefighters did not enter burning buildings to fight fires.

Another reason why firefighters attack fires with portable extinguishers and hand-held hose streams inside burning buildings is because it is safer to attack a fire while it is still manageable. If a fire is allowed to grow and spread outside a building, the dangers to firefighters actually increase (Figure 18.3). When fires spread outside buildings, brick walls collapse, explosions—caused by flammable gases and liquids used for cooking and heating—occur, and nearby high-voltage wires burn and fall from utility poles.

Injuries During Interior Firefighting

Each year, 100,000 firefighters are injured at fires. One of the most dangerous moments on the fireground occurs when a large fire, beyond control of personnel and equipment on the scene, has extended outside a burning building and is spreading rapidly to adjoining exposures (Figure 18.4). At this instant, firefighters, attempting to remove people in the path of the fire and to set up initial attack hose streams, are extremely vulnerable to fireground injury. The five major categories of firefighter injuries—strains, sprains, wounds, cuts, and bruises—often occur at this time. These injuries may be reported after the fire has been extinguished, but they occur during this dangerous period when flames are spreading out of control and firefighters are attempting—perhaps vainly—to prevent fire extension.

Figure 18.3. RIGHT: *If a fire is allowed to increase in size and spread outside a building, the danger to firefighters actually increases. A content fire is less dangerous than a structural fire.*

Figure 18.4. *Some fires are beyond control and extinguishment by interior attack hose lines before the firefighters arrive at the scene. At these fires, where there is a delay in transmitting the alarm, outside defensive attack strategies must be employed.*

Interior Firefighting Equipment

Finally, firefighters go inside burning buildings to attack fires because the equipment they use is designed for interior firefighting. Once flames spread outside a burning building, it is often beyond the capabilities of the firefighters' equipment. Radiant heat spreads fire to nearby buildings and prevents firefighters from approaching the fire. Windblown, burning embers can travel for miles and ignite the rooftops of other buildings. These large fires (conflagrations) are stopped by natural barriers such as rivers and ocean spaces or by man-made barriers such as streets, highways, or solid brick walls. Firefighters move in after the fire spread has been stopped and supervise the controlled burning.

America's fire service is equipped primarily for interior firefighting (Figure 18.5). Two or three firefighters per apparatus, small-diameter (six inches or less) water mains, mobile water supply tankers, small-diameter hose, and SCBA are effective during inside firefighting.

Figure 18.5. *America's fire service is equipped primarily for interior attack firefighting, using small-diameter hose lines.*

Lessons to Be Learned

In 1989, the NFPA tallied 859,500 structural fires in the United States. Only 363, or less than one-half of one percent, resulted in large fire losses ($1 million or more). Based on these statistics, the fire service may logically argue that interior firefighting is successful at more than 99 percent of the fires.

19

Exterior Firefighting Dangers

"VENT THE TOP FLOOR!" orders the chief to the captain of second-arriving ladder company 6 at a fire on a hot, breezy, summer afternoon. The captain and four firefighters climb the aerial ladder that has already been raised to the rooftop of the large, isolated, four-story, 100- by 100-foot brick-and-joist apartment house. Dark brown smoke is pushing out around the six top-floor window frames facing the street front. The fire is located in the rear of the building, on the top floor.

When the firefighters reach the roof and climb from the ladder, windblown smoke immediately engulfs them. Their vision is obscured completely. In the distance, a screeching power saw can be heard on the roof. Unable to see anything in the dense smoke, they crouch down and proceed cautiously toward the sound of the power saw at the rear of the rooftop. With body weight always on the rear leg, each firefighter slides one foot forward, feeling first for an object in front of him, such as a low parapet enclosing a shaft, a skylight, or a scuttle opening. When the forward foot meets no obstruction, it is pressed down to test the roof deck's stability. Only when that stability is assured does the rear supporting leg move up.

Suddenly, the wind and the smoke on the rooftop change direction. The roof of the fire building becomes visible. Two firefighters are in view at the rear of the roof, working feverishly. One is bending at the waist, moving backward, cutting a second vent opening in the roof. Several feet away, another firefighter is pulling up roof decking inside the first vent cut, using the adz end of a halligan tool as a hook. Smoke is rising around the firefighters everywhere from the top-floor fire. Smoke pushes out the vent cut where the firefighter is pulling up roof boards, drifts up around the base of a soil pipe, emerges from the top of an old, sealed-up dumbwaiter shaft roof structure, billows out the open doorway of the bulkhead structure containing the interior stairs, streams from the broken skylight over the stair enclosure, and rolls up over the rear parapet wall across the rooftop.

The firefighters move forward to assist, and the officer orders the power saw started up. As they walk on the roof it feels solid beneath them. The captain glances at the tools carried by his firefighters and assigns tasks. Unable to be heard above the noise of one cutting and one idling power saw, the officer taps the shoulder of the firefighter carrying the axe and the halligan tool and points toward the firefighter struggling with the roof boards. Next he taps the firefighter who is carrying two six-foot pike poles and points toward the rear parapet wall, indicating that the firefighter vent top-floor windows by reaching over the roof edge to break them from above.

The officer and the two remaining firefighters proceed to a point near the firefighters already on the roof. As they approach, the roof deck feels soft and spongy beneath their boots. The captain signals the firefighters to stop. Cautiously, he moves forward, testing the stability of the roof. The solid beams beneath the roof deck can be felt by the officer. Proceeding forward, walking on the part of the roof deck directly over the roof beams, he attempts to find a spot, as close as is safely possible, to vent above the fire below. Although firefighting procedures state that a roof should be vented directly over the fire, the captain knows, by experience at some previous fires, that the roof has been so weakened and burned here that this procedure is too dangerous to attempt. At such a fire, the roof vent must be cut as close as possible to the point directly over the flames.

He is moving close to the original roof cut, out of which flames are shooting up, mixed with smoke. With his body weight on his rear foot, placed above the roof beam, the officer puts the other foot down on the roof deck in front of him, between the joists. The roof deck feels too soft. He retreats several feet to a spot where the roof deck feels more stable, turns to the firefighter holding the power saw, and, with the point of the lock puller he is carrying, draws an imaginary four- by four-foot outline on the roof at the place where he wants the roof vent. As the firefighter lays the high-speed blade on to the roof, the screeching noise of both saws can be heard in the street below. The captain signals the remaining firefighters to act as guides for the roof-cutting firefighters. Each guide places one hand on the back of the firefighter who is bent over, moving backward with the cutting saw. Simultaneously, the guide checks the area behind the firefighter cutting the roof, to ensure that there are no tools, holes, objects, shafts, or low parapet walls that could cause the roof cutter to lose his balance and trip. The continuous contact of the guide's hand on the back of the firefighter with the saw indicates that the area behind him is clear of obstructions.

With the roof vent operation established, the officer surveys the roof. He walks to the bulkhead stair door opening, crouches down, and extends his arm, holding the lock-puller tool, into the smoke- and heat-filled doorway. He slides the tool across the floor of the stair landing, checking to ascertain that no one has collapsed on the floor just inside the doorway, in trying to reach the roof. He proceeds to the side parapet of the roof and checks the progress of the roof venting. Flames shoot straight up out of the first completed roof vent. Firefighters have pulled up the roof boards of the second opening and with pike poles are pushing the ceiling down beneath the roof opening. Holding the hooks at one end, they jab and spear the pointed ends through the ceiling. Smoke, mixed with tongues of flame, rises out of this vent almost into the faces of the firefighters. The third roof cut is being handled by the two firefighters working together with the power saws. Looking over the rear parapet wall, the officer sees a gooseneck ladder leading from a top-floor fire escape to the roof. Thinks he: "When they advance the hose line through the apartment, this will be an avenue of entry to the top-floor apartments to search for victims."

Flames explode out of the top-floor window directly beneath the officer and he pulls his head away to avoid the heat. Looking down the side of the building, the officer sees one firefighter leaning over the edge of the parapet, swinging the pike pole with one hand like a bat, breaking the last window of the burning, top-floor apartment. Flame and smoke are coming out of the row of six broken windows. The officer does not hear the hose stream moving in on the fire below. There is no sound or vibration of the hose stream striking the under side of the roof deck, nor is there water spray shooting out any of the vented top-floor windows. He realizes that something must be wrong downstairs.

Back with the roof cutting team, he observes their good-sized roof vents. Six-foot flames are blowing straight up out of the openings into the afternoon sky; there is very little smoke mixed with the flames. Roof venting accomplished, the firefighters assemble their tools and look toward the officer for new assignments. The captain radios the command post: "Ladder 6 to Battalion 2. The roof is opened and the top floor vented."

"Battalion 2 to ladder 6. Get off the roof! We are having trouble advancing the lines. We are going to hit it with outside streams."

"Ladder 6, ten-four."

There are three reasons why a fire officer orders an outside attack on a fire. These are: firefighters are unable to advance an inside attack hose line; the structure is too dangerous; or there is a shortage of backup resources to respond if the interior attack were to fail.

Heat is the major reason why firefighters are unable to fight a fire using inside hose streams (Figure 19.1). Superheated gases in a confined space bank down to floor level and engulf firefighters advancing a hose line; they are then forced to back out of the area with the hose or temporarily to abandon the hose when the intense heat occurs suddenly.

Heat in a fire area often descends upon firefighters from a concealed roof or attic space above their heads. Fire burning in a con-

Figure 19.1. *An exterior attack on a fire is recommended when the structure is too dangerous to enter.*

Figure 19.2. LEFT: *Firefighters are often burned when attempting to advance an attack hose line into the path of flame and heat.*

Figure 19.3. Obstructions sometimes prevent the success of an interior or exterior hose line.

cealed ceiling space below a tightly sealed roof deck expands the fire gases that cannot rise, so the heat sinks to the floor, where firefighters are trying to advance a hose line. Until the roof or attic space is vented to release the superheated gases upward, firefighters will be prevented from extinguishing the fire by working inside the building.

Wind is another problem that can prevent an inside attack on a fire. Even if the roof is vented, a strong wind blowing through a fire area toward firefighters attempting to advance an attack line will drive heat and flame into their path (Figure 19.2). A fire burning several rooms back inside a fire area cannot be extinguished by a hose stream operated from an entrance door. Only windblown fire gases mixing with air and turning to flame at the entrance door—not the seat of the fire—are cooled by the doorway hose stream. To extinguish any fire with water, it must be discharged directly on the burning material, not on the convection currents. When winds blow at 30 miles per hour or more, a fire officer should anticipate problems with interior firefighting.

Obstructions also prevent advance of an interior attack hose team (Figure 19.3). Room partitions and stock piled up to the ceiling in a store will block a hose stream. Water from a 20- or 30-foot hose stream will be prevented from hitting the fire. In addition to blocking water streams, smoke- and heat-filled rooms, partitioned off into a maze—such as many small rooms or narrow, winding passageways—can make firefighters fearful of advancing too deeply into a fire area. Unless the firefighters are familiar with the room arrangement, they will not move an attack line into a maze-like area. If a flashover occurs, the chance of escape is small. Firefighters become lost in maze-like fire areas and cannot find their way out, even when following the hose line that leads to the outside; excess hose coiled up in several rooms, snaked in and out of the fire area, cannot be used as a guide to get back to safety.

Inability to advance an interior attack hose line is often due to the size of the fire. The fire may be beyond the interior attack extinguishing capabilities of the first response. A fire officer should know the capabilities of the first-alarm response, interior hose line attack.

A rough estimate of the first-alarm response, interior firefighting capability can be calculated. For example, if three engine companies make up the first response, and each company can place one 1¾-inch hose capable of delivering 200 gpm into operation, the inside fire-extinguishing capability of the first response is: The protection of three separate fire areas or of one fire area requiring not more than 600 gpm to extinguish. Roughly, a fire involving four separate fire areas or a fire involving an area that needs 700 gpm for extinguishment is beyond control of these first responders' interior hose line attack.

It is difficult to define accurately the interior firefighting limits of a single fire area. There is, however, a mathematical "fire flow" formula used by the fire service for pre-planning that can estimate fire flow requirements of a single fire area. This fire flow formula, taught at the National Fire Academy in Emmitsburg, Maryland, is:

$$\text{Fire Flow} = \frac{\text{Length of Fire Area} \times \text{Width of Fire Area}}{3} \times \text{\% of Flame of Room or Area}$$

Assuming there are no exposures, by the fire flow formula it can be calculated that a 2,000-square-foot area, or 40- by 50-foot room, with 100 percent flame involvement, requires 660 gpm for extinguishment. Using this formula, a fire area of 2,000 square feet would be considered beyond the interior fire extinguishing capability of three 1¾-inch hose lines delivering 200 gpm. These companies would have to set up outside master streams that deliver more water instead of stretching interior attack hose lines.

There are many fireground factors that influence these fire flow formula calculations: fuel load, layout of fire area, and access to the fire area. The formula does, however, give a fire officer a basis for estimating interior attack fire-extinguishing capabilities of small-diameter hose lines.

Cost-Benefit Analysis

The second consideration that can prompt a fire officer to order an outside defensive firefighting strategy is the danger in the burning building. Target hazards and so-called "fire breeders" are often 100-year-old buildings, abandoned structures, or buildings overloaded with heavy stock. Regardless of the size of the fire or the extinguishing capabilities of the first-alarm responders, a fire officer should consider outside firefighting when encountering a serious fire in an unsafe building.

A fire officer who has a dangerous structure located in his first-response district should conduct a cost-benefit firefighting analysis of the building before a fire occurs. The officers should compare the cost of a disabling injury to a firefighter with the benefit of preserving the dangerous building from fire.

To do so, the officer must determine the worth of the hazardous building. Then the officer should calculate the average age of the firefighter in the fire companies and the amount of disability (if a career firefighter) or workman's compensation (if a volunteer firefighters) he or she would receive if permanently injured. Multiply the annual disability compensation by the number of years that the firefighter would receive payments, up to 70 years of age, the average life expectancy. Then compare the cost of the disabling injury to the monetary benefit of saving the dangerous structure. Consider the following budget bureau view of firefighting:

a. The average replacement cost of a dwelling is ______.
b. The yearly medical pension of disability payment for a firefighter's annual disability pay is ______.
c. If the injured firefighter is 35 years old and lives to age 70, total disability payments will be: Annual disability pay × 35 = Total 35-year cost.
d. Accountant's view of the value of one firefighter's life is ______.
e. The value of a four-man company is ______.

From a fire officer's point of view, there can be no monetary value placed on the worth of a firefighter's life or limb (Figure 19.4). The answer obtained from this crude cost-benefit analysis may be a shock, however. It may cause you to reconsider the risks of interior firefighting. The cost of a disabling injury to a 35-year-old firefighter, using this analysis, is $500,000; that of a four-firefighter company is $2 million. What is the cost of a vacant building? Most occupied buildings in America are not worth $500,000, the cost of a disabled firefighter. Most vacant buildings are worth less than $10,000.

When there are no people inside a burning building, the safety of a firefighter is the highest priority.

Resources

Even when the fire can be extinguished and the structure is safe, there is a situation when the initial hose line interior attack may not be undertaken. This occasion can occur when there are no backup

Figure 19.4. *There can be no monetary value placed on the loss of a firefighter's life or limb.*

Figure 19.5. Firefighters must be trained to use large-caliber master streams effectively.

resources to respond. Large city and suburban fire departments that have almost unlimited backup resources are interior attack-oriented, because, if the interior attack fails, mutual-aid or the backup companies can protect exposures quickly.

Rural fire departments with limited backup resources must be more conservative in their strategy. If the second-arriving company has a response time of 30 minutes or if there is no second-arriving company, firefighting strategy must be defensive.

I adopted this defensive strategy at three large fires one night in 1977 during a power blackout, in the course of which people broke into stores and burned them after the contents had been removed. The communications center notified everyone that there were no backup companies available to respond that night. Arriving at one large fire, I ordered the first hose line stretched into a tower ladder position in front of the burning building. That evening, the importance of backup resources and how they influence interior firefighting crystallized. It showed how dependable backup apparatus response gives confidence to interior firefighting and how undependable backup apparatus response takes it away.

Lessons to Be Learned

Large-caliber outside master streams must be put into operation quickly when needed during an outside attack at a fire; however, this technique is easier said than done. To do it effectively requires preplanning on the fireground. Pumpers must be positioned at water supply sources; apparatus with deck guns or aerial pipes must be places properly near the fire to ensure the effective vertical and horizontal reach of the streams; and, most important, firefighters must be trained to stretch large-diameter hose and to set up and operate complex master stream appliances (Figure 19.5).

20

Safe Firefighting Practices

YOU CANNOT LEARN how to survive the dangers of firefighting by going to college. You may learn most of the important skills you need to be a good firefighter, but you can't learn firefighting survival in school. You can learn it only by working with other experienced firefighters, by observing other firefighters in action, or, unfortunately, by suffering an injury during a fire.

Safe firefighting procedures are passed along from veteran firefighter to rookie firefighter by example at a fire and by conversation and explanation in the firehouse. Safe firefighting techniques are universal. They are the same regardless of where you fight fires. Building construction and firefighting procedures may vary, but safe operating procedures on the fireground are universal.

The rest of this chapter offers safety practices for some of the most dangerous firefighting operations, techniques that are known and sometimes taken for granted by veteran firefighters but are unknown to young recruits. All firefighters should understand and use these procedures.

Above-the-Fire Operations

1. When stretching a hose line to an upper floor of a building, do not pass a floor on fire unless a charged hose line is in position on that floor.
2. Notify your officer when going above a fire to search for victims or vertical extension of flame or smoke (Figure 20.1).

Figure 20.1. *Notify your officer when you go above a fire to search.*

3. When climbing or descending a stairway between the fire floor and the floor above, stay close to and face the wall. Heat, smoke, and flame rise vertically up the stairwell.
4. If you enter a smoke- and heat-filled room, hallway, or apartment above a fire and suspect flashover conditions behind you, locate a second exit—a window leading to a fire escape or a portable ladder, for example—before initiating the search.

Advancing a Hose Line

1. Crouch down and keep one leg outstretched in front of you when advancing an attack hose line in a smoke-filled fire room. Proceed slowly, supporting your body weight with your rear leg. Your outstretched leg will feel any hole or opening in the floor deck in your path of advance.

2. To prevent getting driven off a fire floor by rollover (the sudden ignition of combustible gases at ceiling level) while waiting for the hose line to be charged, crouch down outside the burning room or apartment and, if possible, close the door to the burning area. When the line is charged, open the door and immediately attack the fire.
3. During a fire in a one-story strip store, vent the roof skylight over the fire before advancing the hose line in order to prevent injury from backdraft, explosion, or flashover.
4. When it is not possible to vent the rear or the roof of a burning store quickly and signs of backdraft or explosion are evident from the front of the store, vent the front plate-glass windows and doors, stand to one side, let the superheated combustible gases ignite temporarily, and then advance the hose line for fire attack (Figure 20.2).

Figure 20.2. *When a smoke explosion is possible and roof venting or rear venting cannot be accomplished, firefighters should vent the plate-glass windows, stand to one side, and let the fire gases explode or burst into flame. Then they should move in and extinguish the blaze.*

Cellar Fires

1. Self-contained breathing apparatus must be worn before entering the cellar of a burning building, even if there is only a light haze of smoke (Figure 20.3). Carbon monoxide, a deadly, gaseous by-product of combustion, is colorless, odorless, explosive, and easily trapped in unventilated areas.
2. Notify your officer and wear self-contained breathing apparatus before entering a cellar to shut off utilities. If the officer doesn't receive confirmation of the shut-off within a reasonable amount of time or is unable to establish contact, he must make an immediate effort to locate the firefighter and assure his safety.
3. Do not let the presence of an operating sprinkler give you a false sense of security. Wear your SCBA before entering. Carbon monoxide gas can be present even when a sprinkler is discharging and controlling a smoldering fire.

Collapse Rescue

1. At any collapse, stretch a hose line and charge it to protect possible victims and rescuers from sudden explosion and flash fire (Figure 20.4).
2. Shut off all utilities—gas, electric, and water—immediately upon arrival at a building collapse. Do not wait for the utility company.
3. Heavy mechanical equipment, such as cranes and bulldozers, should never be ordered to remove collapsed portions of a building while hand digging is being done.
4. Parts of a structure that are in danger of collapsing, during a rescue operation should be shored up, never pulled down.

Fire Escape Operations

1. When climbing a fire escape during a fire, always hold, with one hand, a part of the fire escape itself to prevent serious injury should a stair tread suddenly give way.
2. Before climbing a gooseneck ladder leading from a top-floor fire escape landing to the roof, vigorously pull the ladder away from the building in order to test its stability. If the metal fire

Figure 20.3. TOP: *Masks must be donned before entering the cellar even if there is only a light haze of smoke. Carbon monoxide accumulation often occurs in below-grade areas.*

Figure 20.4. *A hose line should be stretched and charged when you are operating at a building collapse, in order to protect buried victims and rescuers from gas explosions.*

escape or the wooden or masonry structure to which it is attached is corroded, the gooseneck ladder could pull away from the building.

3. When taking up after a fire, the fire escape drop ladder is to be returned to and secured to its normal raised position. Firefighters should never attempt to descend to the street from the first fire escape balcony by climbing down the drop ladder in raised position and then dropping down from it to the sidewalk. Pendulum hooks holding fire escape drop ladders have been known to break suddenly from their connections, and firefighters on them have been seriously injured. A firefighter should use a portable ladder or enter an apartment served by the balcony in order to descend to the street level.
4. Stand away from the weights when lowering a counterbalance weighted ladder. The weights may collapse from the impact of the ladder striking the sidewalk (Figure 20.5).

Forcible Entry Operations

1. When forcible entry is required for an inward-swinging door behind which there is intense heat and fire, the inward swing must be controlled. A firefighter or an officer should hold the doorknob closed with a gloved hand or a short piece of rope while other firefighters force the lock open.
2. A firefighter performing forcible entry on a door to an apartment on fire is extremely vulnerable to injury from backdraft or smoke explosion once the door has been opened and air flows into the fire area. The firefighter who believes he can avoid a blast by observing warning signs or by reacting in a split second is in error—explosions happen too fast. The only real protection a firefighter has against explosion is his protective equipment—gloves, mask facepiece, helmet, earflaps, turnout coat, pants, and boots—properly arranged and in good condition.
3. Generally, when a firefighter must use an axe for entry, it should be moved forward forcefully in a punching action (Figure 20.6). The power behind the axe movement comes from the firefighter's shoulder and the weight of the axe, not from the swing. If it is necessary to swing an axe during a forcible entry operation, first check for the presence of nearby firefighters and overhead obstructions.

Figure 20.5. *TOP: Firefighters should stand clear of metal counterbalance weights when lowering fire escape stairways.*

Figure 20.6. *Swing an axe forward forcefully in a punching action to maintain safe control of the tool.*

Figure 20.7. Ladder belts or safety harness should be used when you are climbing or operating on an aerial ladder.

Ladder-Climbing Operations

1. A firefighter ordered into a room from outside via a ladder should first place any tools he is carrying on the floor inside the window before entering. Then, with both hands free, he should grasp a portion of the window and test its stability. If it does not move, the firefighter will maintain his grip on the window while moving through it from the ladder.
2. A firefighter climbing an aerial ladder should use a ladder belt to secure himself to the rungs (Figure 20.7). A leg lock should not be used as a substitute for a ladder belt.
3. Firefighters should never be up on an aerial ladder while it is being raised, rotated, or extended. The ladder must be in position before anyone climbs it—that means making sure that the ladder locks are set, too.
4. The degrees of safety afforded in removing a victim from a burning building are, from highest to lowest: smokeproof tower, interior enclosed stairway, safe fire escape, aerial platform, and aerial ladder.

5. When climbing into the window of a burned-out or vacant building, drop your tools inside the window before entering and listen to them strike the floor. If you don't hear the tools strike the floor, either the window opens into an elevator shaftway or the floor has burned away.

Wall Collapse Dangers

1. Whenever there is a danger of wall collapse, an officer in command must establish a collapse danger zone. A collapse danger zone should be equal to the height of the unstable wall. All firefighters should be withdrawn from the burning building to a distance at least equal to the height of the wall.
2. The officer establishing the collapse danger zone must take into account not only how far outward the wall may collapse but also the horizontal span of possible wall collapse.
3. The collapse danger zone for an aerial stream will vary from that established for ground stream operations. The apparatus should be parked outside the collapse danger zone, and the raised aerial ladder or platform should never be placed inside the curving path of a wall collapsing at a 90-degree angle (Figure 20.8).
4. Establishing a collapse zone for tall structures could require firefighters to be positioned beyond the reach of hose streams. In this case, a "flanking" position is called for: The master streams must be placed in front of the adjoining buildings or at corner safe areas of the fireground. The master stream's range and effectiveness will be reduced, but the life safety of the firefighters will be ensured even if the unstable wall falls outward.
5. The greatest danger of wall collapse occurs during an outside attack on a fully involved church fire. There are only four safe areas in which to park vehicles and operate master streams at such an incident: the four corner safe areas (Figure 20.9). If all of the walls collapsed outward simultaneously (however unlikely), only these four areas would be safe from falling debris.

Overhauling

1. After a fire has been extinguished and before overhauling begins, three safety actions should be ordered by the officer in

Figure 20.8. TOP LEFT: *Aerial ladders or platforms should not be placed inside the curving path of a possible wall collapse when there is a danger of structural failure at a fire.*

Figure 20.9. TOP RIGHT: *At a fully involved church or mill building, master streams or aerial platform apparatus should be positioned in the corner safe areas to avoid possible collapsing walls.*

Figure 20.10. *Every firefighter should always carry a personal flashlight.*

command: Fresh air should be pumped into the hot, smoke-filled area by fans or by the ventilation system; portable lights should be set up to improve visibility; and a safety survey of the structure and contents should be undertaken, checking especially for collapse hazards, hazardous materials, and utility shut-off control valves.

2. Firefighters ordered to shut off utility control valves for gas or electric power must consider the possibility of carbon monoxide and smoke accumulation in the cellar, particularly when a fire of long duration has been extinguished in a first-floor store directly above the cellar and the cellar is completely below grade and without windows. Self-contained breathing apparatus must be worn.
3. The firefighter's best protection against injury and death from a fall during overhauling is a properly charged flashlight (Figure 20.10). No firefighter should respond to a fire without a personal light.
4. The most potentially dangerous area of local floor collapse inside a burned-out residence is the bathroom. The weight of a firefighter is enough to trigger the collapse of a fire-damaged bathroom floor.
5. If flames are discovered still burning at a gas meter or broken pipe after a fire has been knocked down, do not extinguish the flame. Let the fire burn, protect the exposures with a hose stream, and alert command that the gas has to be shut off at the cellar or street control valve.

Propane Cylinder Fires

1. Full protective clothing—including mask facepiece—must be in place before a firefighter approaches a 20-pound propane cylinder to shut off the control valve when a small flame is burning at an outlet. There is a danger that the relief valve will suddenly activate, creating a fireball that could engulf the firefighter.
2. To protect a propane cylinder from exposure to a nearby fire, direct the hose stream to the top portion of the tank. This top portion contains vapor; it is in this vapor space that most propane cylinders BLEVE, due to heat from fire exposure.
3. When a propane cylinder is discovered burning around the cylinder valve, employ the following tactics: Cool the vapor space. After the area has been cooled with water for ten min-

utes and the flames appear stabilized in size and intensity, approach the valve, wearing full protective equipment and mask, and shut off the gas by means of the control valve, if possible. If the flow of burning gas cannot be shut off, allow the propane cylinder to burn itself out, and use the hose stream to protect the exposure.

4. The firefighter shutting off the flow of burning gas at the propane cylinder outlet should be protected by a wide-pattern, low-velocity stream; position the fog stream between the control valve and the burning outlet. The firefighter's hand should be behind the fog curtain when turning the control valve; the flaming outlet should be in front of the fog curtain.
5. If you are in doubt about how to control a fire involving a propane cylinder, move all civilians and firefighters to a safe distance beyond the explosion danger zone, get behind a barrier, and let it explode.

Responding to and Returning from Alarms

1. The safest riding position on a fire apparatus is inside an enclosed cab, seated, strapped in with a seat belt. The most dangerous riding position is the tailboard (Figure 20.11).
2. Never turn your back to oncoming traffic. When stopping traffic, use a flashlight, face the oncoming vehicles, and be ready to jump out of the way if they do not stop.
3. The most dangerous side of a fire apparatus making a turn is the "inside of the turn." If a firefighter attempts to jump on to the turning vehicle and misses the side step "inside the turn," he will be run over by the back wheels.
4. Defensive responding means knowing that the entire company is responsible for the arrival of the apparatus at the scene of a fire, not just the driver and the officer.
5. When a fire company arrives on the scene of a highway emergency and there are no police present to control traffice, firefighters themselves must first control the oncoming traffic before safely turning their attention to the emergency.

Roof Operations

1. When walking on a peaked roof, straddle or stay near the ridge rafter. If you slip or lose your balance you can grab the roof peak—it's your one true handhold (Figure 20.12). Chimneys,

Figure 20.11. RIGHT: *Firefighters should not ride the tailboard of a responding fire apparatus.*

Figure 20.12. *Firefighters should straddle or stay near the ridge rafter of a peaked roof. If they lose their footing, the ridge rafter can be grabbed, which may prevent falling from the roof.*

television antennas, and soil pipes are not designed to support the weight of a falling firefighter and may break.

2. To maintain footing when walking on a peaked-roof surface, bend your legs at the knees and walk flat-footed. This method is called the "roofers' walk." It will reduce your chances of sliding down a peaked roof.
3. When there is a danger of peaked-roof deck burn-through or collapse due to an attic fire, place a roof ladder on the sloping side of the roof from which you're operating and walk on the rungs of the ladder. The ladder should be supported by the roof ridge and the bearing walls of the house.
4. Roof operations should be conducted from an aerial ladder or an aerial platform when peaked-roof beams are in danger of collapse due to fire destruction of the attic. The firefighters should, therefore, be independently supported.
5. Firefighters should not walk on a peaked roof with a slope of more than a 30-degree angle from the horizontal.

Searching for the Location of a Fire

1. To reduce your chances of being severely injured by flashover during a search, practice a safe, organized search method. Most firefighters killed by flashover are disoriented and lost in smoke. When searching a small room, maintain contact with a wall and move consistently in a clockwise or counterclockwise direction. In a large or complex area, use a search rope as a guide.
2. With the increasing use of lexan windows, sliding scissor gates, and bars on windows, firefighters searching for the location of the blaze or for victims should always return to the entrance door (Figure 20.13). If a firefighter passes the fire and carries a victim to a fire escape window, they could both be trapped. Crime, or the fear of it, prompts many residents to lock up the second exit.
3. Firefighters should know the warning signs of flashover. When smoke and superheated gases force you to crouch down to less than half the height of the room, there is danger of flashover. Rollover is also a sign of possible flashover—when flashes of flame at the upper part of a room or at the top of a door or window mix with the smoke and heat and flow out of the opening. When you suspect flashover, raise a hand above your head to check for heat build-up.

Figure 20.13. Firefighters cannot depend on a window as a second exit. Lexan unbreakable window glass or window gates may block the window opening.

4. Firefighters should know why the flashover phenomenon has become more common in recent times:
   - Because of the use of smoke detectors, firefighters are arriving at the scene earlier in the growth process of the fire, frequently before flashover.
   - The synthetic furnishings of a typical home are petrochemical derivatives that accelerate flashover by liberating greater amounts of heat and flammable gases.
   - The improved quality of protective gear and equipment has allowed firefighters to enter farther into superheated atmospheres prior to flashover.
   - Thermal windows and energy-efficient heat barriers behind walls and ceilings of rooms hold more heat within the confined space of the room.

Figure 20.14. *The perimeter of a burning building is a dangerous area because of possible falling objects.*

Outside Venting

1. When operating around the perimeter of a burning building, a firefighter performing outside venting must take precautions to avoid injury from falling objects (Figure 20.14). When you hear glass breaking, don't look up. Size up the venting assignment from a distance. Choose the window you want to vent, move in close, vent it, and back away from the structure.
2. To determine the proper angle for placing a ground ladder, stand erect at the base of the ladder with your boots against the ladder beams and your outstretched arms grasping the rungs at shoulder level. If you can do this, the ladder is at the proper climbing angle.
3. When you cannot open a window manually to vent smoke from a building and must break the glass, stand to one side (if possible, the windward side), use a six- or eight-foot pike pole for safe reach, strike the glass with the pike pole at the topmost

area of the window, and work downward. If there is a possibility that firefighters are searching inside the room, first tap the window and only break a small portion of the glass—this action will serve as a warning. Then remove the entire window with the tool. Keep your helmet eyeshield down for protection, wear gloves to protect your hands, and don't stand in front of the window.

4. After flashover occurs inside a superheated, smoke-filled room, there is a point of no return beyond which a firefighter cannot escape back to safety. The point of no return, or maximum distance that a firefighter can crawl inside a superheated room and still get back out alive after flashover, is five feet.

Wildfire Operations

1. When moving through brush during a fire, the firefighter should raise a tool or arm in front of his face as he moves forward to avoid injury from shrubbery, pointed conifer needles, sharp-edged leaves, or abrasive vines. Firefighters walking behind the lead firefighter should space themselves several feet apart in order to avoid whipping branches or leaves.
2. Never enter cattails or brush that is over your head and reduces your vision. If the wind changes, you are in danger of being engulfed by fire in the brush (Figure 20.15).

Figure 20.15. *During a wildfire, firefighters should never walk into brush that reduces visibility. A wind change could encircle you with flame.*

3. When the wind frequently changes direction during a brush-fire operation, the safest location from which to attack the fire is the blackened, burned-out area.
4. A survey revealed that firefighters are most often killed and injured at small brush fires in isolated portions of larger fires; they are not killed by large timberland forest fires. Firefighters are burned to death trying to outrun brush fires, or they are engulfed in flames when a brush fire suddenly flares up around them. Firefighters should attack a brush fire from the flanks—the sides of the fire area that are located between the head (the edge along which the fire is advancing) and the rear.
5. The three most common injuries to firefighters during brush firefighting are eye injuries, falls, and heat exhaustion. Eye-shields must be worn. Firefighters should walk on roads or well-traveled paths whenever possible.

   To lessen the possibility of exhaustion on hot days, rest periods, an adequate water supply, and the removal of protective clothing when safety permits are in order.

21

Firefighting Dangers

Causes of Death and Injury

WHAT IS A falling object? What exactly do the experts mean when they say "in contact with" is one of the second leading causes of firefighter death? What are products of combustion? These statistical categories of death have little significance to a firefighter; they are meaningless terms used by statisticians and administrators who compile death and injury surveys of the fire service. The catagories only serve the "bean counter's" need to fit our firefighting tragedies neatly into a single column. The following are fifty understandable, comprehensible causes of firefighter death and injury. Read them and mourn!

Auto-exposure is the spread of flames from one floor to the floor above on the outside of the building. Flames can be sucked up to the floor above the fire from window to window (Figure 21.1). Firefighters entering a window on the floor above a fire from a ladder or a fire escape can find their escape path back to the window cut off by auto-exposure flame spread.

A backdraft is an explosion caused by the rapid ignition of fire gases occurring in a burning room that has been tightly sealed up. The trigger for a backdraft explosion is the fresh air which enters during the initial search and entry operation of firefighters (Figure 21.2). Firefighters performing forcible entry operations are sometimes killed or injured by the blast of a backdraft. The tightly sealed-up burning room contains combustion gases and high temperatures; most of the

Figure 21.1. LEFT: *Auto-exposure is the vertical spread of fire from one floor to the floor above, by flame spread from window to window.*

Figure 21.2. *The triggering effect for a backdraft explosion is fresh air, which enters with firefighters during their initial search and entry.*

oxygen has been consumed by the smoldering fire. When a door to the superheated room is opened, air is introduced and completes the fire triangle necessary for a sudden rapid explosion.

A blasting agent is an explosive material widely used at construction sites for demolition. This material consists primarily of ammonium nitrate and a fuel, such as number two fuel oil. The danger of a blasting agent is often underestimated when compared with other explosives; this belief is a deadly error in judgment. A blasting agent requires a stronger heat or shock source for detonation than does a high explosive such as dynamite; however, when a blasting agent explodes, it is just as powerful as a high explosive. Firefighters must realize that a blasting agent is just as dangerous as dynamite and must be treated with the same caution. Moreover, the flames of a fire serve as a commonly occurring detonator, which often causes the explosion of a blasting agent. Six Kansas City firefighters were killed when a blasting agent exploded during a fire in a truck.

A BLEVE is a *b*oiling *l*iquid *e*xpanding *v*apor *e*xplosion that occurs when a container of liquefied petroleum gas ruptures. A "BLEVE" can kill firefighters in the fireball created by the ignition of the suddenly released vaporizing liquids, by the rocketing pieces of steel shrapnel flying through the air, and by the shock wave or the blast which occurs during the explosion. Deaths from burns have killed firefighters 250 feet away from large liquefied petroleum containers. Deaths caused by flying pieces of shrapnel from large exploding containers have occurred 800 feet away from the source of the explosion. Five firefighters from Buffalo, New York, were killed when a building collapsed on top of them as a result of the explosion of a leaking propane gas cylinder.

A boil-over is a sudden eruption of hot oil over the top of a large, burning crude oil storage tank. A boil-over could occur after water from hose streams sinks to the bottom of the burning oil and is heated to its boiling temperature. As the water turns to steam and expands 1,700 times, it can cause a boil-over, which could spray boiling oil over firefighters operating hose lines near burning oil tanks.

Carbon monoxide (CO) gas is a toxic product of combustion. During a structural fire, there is usually insufficient oxygen for complete combustion to take place. The uncontrolled smoldering of a fire generates carbon monoxide. There may be gases in a fire area more toxic than carbon monoxide; however, carbon monoxide is produced in large quantities. It is colorless, odorless, and explosive. When mixed with air at low concentrations, 10,000 parts per million of carbon monoxide can cause death when inhaled for one minute (Figure 21.3).

A cellar is a below-grade floor level in a building. Firefighters die in cellars as a result of carbon monoxide accumulation because of incomplete combustion; they die because the oxygen in the air is depleted due to a flash fire; they drown in water-filled cellars; they die in cellars when heavier-than-air gases accumulate there; and they are killed when cellars filled with combustible gases explode during fire. Some cellars are more dangerous than others. A cellar that is completely below grade, without any windows to the outer air, is more dangerous than a cellar that is only partially below grade or one that has windows which can provide ventilation. Cellars in high-rise buildings do not have windows. Subcellars, which are located

Figure 21.3. *Carbon monoxide, when mixed with air at 10,000 ppm, can cause death when inhaled for one minute.*

Figure 21.4. *Carbon monoxide build-up in cellars occurs more readily than in above-grade building fires.*

directly below cellars, are the most dangerous type of below-grade area; they contain no windows and are two stories below street level. A cellar becomes still more dangerous after the fire has been extinguished: smoldering embers generate carbon monoxide (Figure 21.4). Masks should be worn during overhauling in cellars due to the danger of carbon monoxide build-up.

A collapse danger zone is the most deadly area on the fireground. A firefighter should never enter a collapse danger zone after it has been declared. Such a zone is defined as the ground area that a falling wall will cover with bricks or other wall materials during a collapse. The zone occupies that distance from the foot of the unstable wall and extending out for a distance equal to the height of the wall

Figure 21.5. The area inside a collapse danger zone will be completely covered with tons of masonry during a wall collapse.

(Figure 21.5). When a brick or wood wall collapses in a 90-degree-angle collapse, it will kill any firefighter operating near the wall within the collapse danger zone. For example, a wall 20 feet high, collapsing at a 90-degree angle, will kill firefighters operating closer than 20 feet away from the wall.

Collapsing structure is defined as any portion of a burning structure that collapses due to fire damage. Structural collapse is the fourth leading cause of firefighter death (number 1 is stress; number 2 falls, falling objects, in contact with; number 3 is products of combustion). Firefighters are killed by structural collapse outside burning buildings as well as inside them. Unlike the other leading causes of firefighter death, when a building collapses during a fire, large numbers of firefighters die in a single event. Chicago lost 21 firefighters at a single structural collapse during a fire. Philadelphia lost 14 firefighters, New York lost 12, and Boston nine. Beware of building collapse.

A commercial building fire in a store, office, or warehouse is more dangerous than one in a residence building. Most firefighters are killed and injured fighting fires in residence buildings, because of the large number of fires that occur in residence structures. When we examine the number of firefighter deaths per fire incident, however, we find that per fire incident, more firefighters are killed in commercial building fires. When responding to a fire in a commercial building, firefighters should take extra precautions. There are additional dangers present in a commercial building which are not found in a residence building, such as dangerous industrial processes using chemicals and flammable liquids, dangerous machinery, unusual floor layouts, heavier floor loads, larger floor areas, high ceilings, and greater fuel loads (Figure 21.6).

Figure 21.6. *A store or commercial occupancy is more dangerous than a residential occupancy. It has greater fuel load, more likelihood of flammable liquid storage, unusual floor layout, and larger areas.*

Convection currents, the transmission of heat upward through flame and heated smoke, are a dangerous type of heat transfer at a structural fire. Convection currents trap and kill firefighters operating both on the floor above a fire and in cellars (Figure 21.7). Firefighters searching the floor above a fire can be cut off by flame and superheated smoke or gases flowing up an interior stairway; they will be unable to retreat back down stairs filled with the rising convection currents of heat from the fire below. Firefighters operating below grade in a cellar, crouching down, battling a stubborn blaze for a long period of time, may not detect the heat and flame building up over their heads. If the convection currents of heat and flame flow out of the cellar, up the stairs to the street level, and suddenly ignite or fill the stairway, their presence could trap the firefighters in the cellar. Unable to climb up the stairway that is filled with flame or heat carried by these convection currents, the firefighters will die in the cellar.

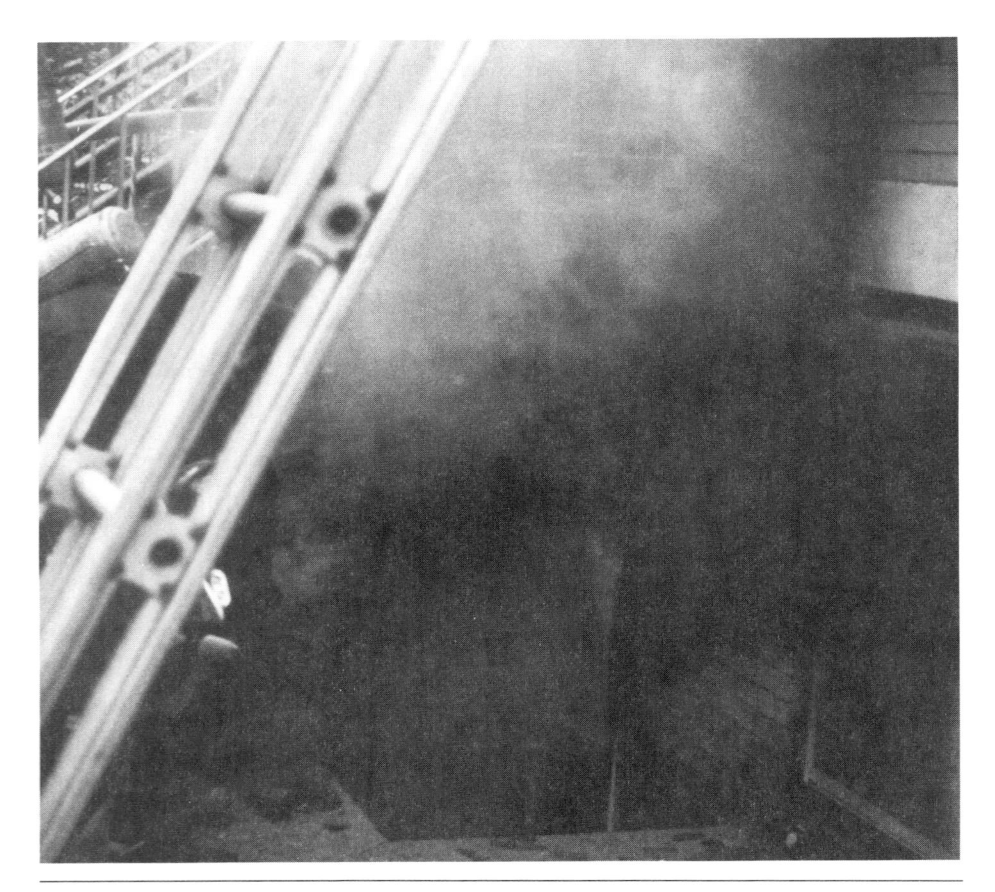

Figure 21.7. *Convection currents of flame and heat trap firefighters operating in a cellar or on the floor above the fire.*

Figure 21.8. TOP: *Disorientation and becoming lost in smoke are the real killers of firefighters trapped in flashover.*

Figure 21.9. BOTTOM: *Sudden flashover is the most dangerous flash phenomenon occurring during a fire.*

Disorientation is the loss of a sense of direction experienced by firefighters when searching in a smoke-filled room. Disorientation is caused by the obscuring smoke and the failure to use an organized search technique when moving around such a room. The condition prevents firefighters from returning to the safety of the door or window through which they entered the burning room. Firefighters who become disoriented (Figure 21.8) and lose their sense of direction in a smoke- and heat-filled room are often subsequently killed by flashover (Figure 21.9) or die from asphyxiation in smoke after their tanks run out of air. Firefighters have died in small, smoke-filled rooms, their bodies found next to doors or windows in a ten- by ten-foot room.

Electric shock is deadly when electric current is transmitted through some part of a firefighter's body. Electricity passing through the body may bring about death by causing violent muscular contractions of the heart, by interrupting the breathing process, or by burning a vital internal organ in the path of the electric current through the body. Most firefighters who are killed or injured by electrocution come in contact with overhead utility wires when climbing ladders or operating in the buckets of aerial platforms. All electric wires and equipment should be considered live and dangerous.

An elevator is a vertical transportation system in high-rise buildings; it must be considered a death trap during a fire in such a building. An elevator can kill firefighters in these ways: *(1)* The elevator may be called to a floor fully involved with fire when the elevator call button is in contact with flame, heat, or smoke. Firefighters in that elevator may be carried to an upper floor of a high-rise building, where the doors suddenly open directly into a flaming lobby. *(2)* In other fires, an elevator may unexpectedly stop between floors while carrying firefighters. If the elevator becomes stuck between floors located above the fire, the elevator shaftway may become a chimney flue, and the firefighters trapped in the car will die from the heat and smoke rising up the elevator shaftway. *(3)* Firefighters have been killed by elevators in high-rise buildings by falling down the elevator shaftway. A firefighter searching in a smoke-filled hallway could walk into an open elevator doorway and fall down the shaft.

An explosion is a violent combustion reaction of fuel, oxygen, and heat which creates rapid expansion of gases strong enough to collapse an enclosing structure or to create shock waves that break glass windows or knock down firefighters near the explosion. There are many types of fireground explosion; a BLEVE, a backdraft, and a smoke explosion are all classified as explosions. A flashover is not an explosion, but the rapid ignition of fuel, oxygen, and heat, which creates rapid expansion of fire gases but is not strong enough to collapse a building, break a glass window, or knock down a firefighter. If a flashover were to cause one of these destructive events it would not be classified as a flashover; it would be an explosion.

Explosives are materials, usually solids, that undergo explosive decomposition when subjected to shock, heat, or pressure. There is a definite possibility of an explosion when these materials are exposed to fire. Explosives are classified under the following categories:

Primary high explosives	(Example: mercury fulminate); Mild shock or heat detonate primary high explosive.
Secondary high explosives	(Example: dynamite and nitroglycerin); More powerful than primary high explosives, secondary high explosives are detonated by shock from a primary explosive.
Low explosive	(Example: black powder, smokeless powder, and rocket fuels); Fire constitutes the greatest hazard to low explosives.

The Department of Transportation divides explosives into four main classifications for transportation purposes.

Classification	*Hazard*
Class A explosives	Maximum hazard; includes dynamite, nitroglycerin, mercury fulminate, black powder, and blasting caps.
Class B explosives	A high flammable hazard; includes most propellant materials.
Class C explosives	A violent explosive hazard; includes fireworks, explosive rivets, and detonating cord.
Blasting agents	When exploding, these are similar to a class A explosive; includes ammonium nitrate.

Firefighting should never be attempted when the flames have reached a primary high explosive, a secondary high explosive, a low explosive, a class A, B, or C explosive, or a blasting agent. Firefighting should be stopped and firefighters withdrawn rapidly to a minimum distance of 2,000 feet from the burning explosive.

Falling is one of the second leading causes of firefighter death. (*Falls, falling objects*, and *in contact with* constitute the second leading category of firefighter death.) (See Figure 21.10.) The most deadly falls suffered by firefighters occur from elevations: falls from responding fire apparatus, falls from roofs of burning buildings, and falls from fire department ladders; however, most injuries from falls on the fireground occur at ground level. Firefighters trip over objects in the dark during fire operations, they fall through burned-out floors, they

Figure 21.10. *Falls, falling objects, and contact with electricity are the second leading fireground causes of death.*

Figure 21.11. *Flame temperatures vary between 2,500 and 3,500 degrees F.*

are knocked downstairs by other firefighters attempting to escape flashover or explosions, and they slip on ice- or snow-covered steps or sidewalks when carrying tools.

A falling object is any material that falls from, is thrown out of, or breaks off a building or any nearby structure during a fire. Some examples of falling objects which kill and injure firefighters are smoldering furniture thrown out of windows during overhauling, falling tools which have slipped out of the hands of firefighters, windows that have been vented from inside a burning building, and people jumping from buildings to escape flames. The most dangerous area on the fireground is that around the immediate perimeter of a building. To escape the danger of falling objects (but not of collapse), get inside the building or stay away from the perimeter of the burning building.

Flame is the luminous zone of combustion when one gas burns in another. Flame temperatures vary between 2,500 and 3,500 degrees Fahrenheit (Figure 21.11). Thermal burns to firefighters are one of the third categories of firefighter death (*flame, gases, heat, and smoke—products of combustion*—are the third leading cause of fireground death). A firefighter's best protection against flame is water from an attack hose stream. The insulation of protective firefighting gear and mask will protect a firefighter from more serious injury when exposed to the flame of a flashover, reflash fire, or flash fire. Nothing can protect a firefighter from prolonged exposure to flame. Flame is the most deadly and most common "hazardous material" a firefighter will ever encounter.

A flameover is a flash fire which occurs over the surface of a wall, ceiling, or floor. It is caused by the sudden ignition of combustible vapors that are produced by heating the surface of that wall, ceiling, or floor (Figure 21.12). Flameover fires can be caused by the combustible surface coating of polyurethane or some other flammable finish. When the combustible coating over a wall or ceiling is heated by a nearby fire, it creates a flammable vapor mixed with smoke, which will suddenly flash into a surface fire. Flameover fires trap firefighters searching for fires and advancing hose lines down hallways. Wood-panelled walls, school desks, theater scenery, and decorative wall and ceiling coverings can cause a flameover fire.

A flammable-vapor explosion is caused by the instant ignition of flammable vapors and gases mixed in air and set off by a source of heat. Flammable-vapor explosions often occur during fires started by an arsonist, who uses a flammable liquid to hasten the spread of a fire. Unexplained explosions which occur during fires are often flammable-vapor explosions. When an explosion or flash fire occurs in a room or occupancy adjoining the area of fire origin, a flammable-vapor explosion, caused by a flammable liquid used by an arsonist to start the fire, should be suspected. Flammable-vapor explosions often occur in areas adjoining a fire. The flammable vapors, from the flammable liquid used to start the fire, drift into the adjoining occupancy and may explode even after the main fire has been extinguished; sparks from the main fire may ignite the flammable vapors which have spread to the adjoining occupancies.

Figure 21.12. *Flameover is the sudden ignition and rapid fire spread along the surface of a heated wall or ceiling.*

A flare-up is the sudden explosive flaming of a brush fire caused by a strong wind gust or change in wind direction (Figure 21.13). Firefighters working in high, dense brush, such as cattails or chaparrel, have been trapped and killed by flare-ups when fighting wildfires.

A flash fire is the sudden rapid ignition and the immediate self-extinguishment of a room filled with a combustible atmosphere. Flash fires are caused by pockets of combustible gases, vapors, or dusts which suddenly come in contact with a nearby ignition source. Because the flammable vapor, gas, or finely divided dust is at or below its lower limit of flammability, there is insufficient combustible atmosphere to sustain continued burning; the fire extinguishes itself.

A flashover is defined as the rapid ignition of heated fire gases and smoke that have built up in a burning room. Flashover is caused by thermal radiation feedback, sometimes called re-radiation, from the ceilings and upper walls, which have been heated by the fire growing in the room. When all the combustibles in the space have been heated to their ignition temperatures, simultaneous ignition of the room occurs. Flashover means full room involvement with fire; it takes place during the growth stage of a fire (Figure 21.14). After flashover occurs, firefighters and civilians in the room are killed, all searching stops because the fire has become too severe, an attack hose line is now required for extinguishment, and there is a possibility of collapse.

A hazardous material is any substance that can cause death or disabling injury in a brief exposure. The most common hazardous materials that a firefighter will encounter are products of combustion: flame, heat, toxic gases, and smoke. Fire kills more firefighters than any other known hazardous material.

The head of a wildfire is the fast-moving leading edge along which a grass fire, brush fire, or treetop fire (crown fire) is advancing. The head of a wildfire is the most dangerous area of the fireground; firefighters are trapped and killed there by the rapid spread of flame. You may find the head of a wildfire at the leeward side of a large-area fire or advancing up a slope or mountainside.

Heat is a product of combustion. A fire will produce flame, heat, smoke, and toxic gases. Heat is associated with the natural motion of molecules; the faster the molecules of a material move, the hotter the

Figure 21.13. *Flare-up is the sudden explosive flaming of brush.*

Figure 21.14. *Flashover takes place during the growth stage of a fire.*

Figure 21.15. *A high ceiling in a commercial building can allow dangerous heat and flame build-up to take place over the heads of firefighters.*

material becomes. A firefighter's protective clothing and breathing equipment cannot protect him from the heat of a fire. Dry air temperatures above 280 or 320 degrees F will cause extreme pain to unprotected skin. Death by hyperthermia will occur if a body absorbs heat faster than it can be dissipated by the evaporation of surface moisture.

High ceilings more than ten feet above floor level constitute a danger to firefighters. Such a ceiling in a commercial building can allow dangerous heat and flame build-up to take place above the heads of firefighters searching in smoke (Figure 21.15). In a smoke-filled room, a firefighter sizes up the flashover danger by how low he must crouch to crawl under the heat banking down from the ceiling. Most rooms in residence buildings have ceilings eight or ten feet above floor level. When a firefighter searches a smoke-filled room in a commercial building whose ceilings are 15 or 20 feet above floor level, that firefighter can make a serious error in judgment; his size-up of the potential flashover fire spread and the heat build-up over his head can be dangerously incorrect.

A large-area occupancy is an enclosure greater in size than 25 by 50 feet with no interior enclosing partitions. The open interior design of large-area occupancies, such as warehouses, places of assembly, and stores, causes firefighters to become disoriented in smoke. Search and rescue in a smoke-filled large-area occupancy is more dangerous because of the possibility of becoming lost in dense smoke, unable to find the way safely back to the entrance, and, after the SCBA runs out of air, being asphyxiated or caught in a flashover. A search rope should be used by firefighters searching in a large-area occupancy, with one end of the rope tied near the entrance door and the other end played out as they search the interior of the occupancy. The purpose of the search ropes is to guide the firefighters back to safety when visibility is reduced by heavy smoke.

A master stream is a ground-based or aerial nozzle with a fog or straight stream capable of delivering more than 300 gallons of water per minute to a fire (Figure 21.16). If a master stream is delivering three or four tons of water through a straight stream nozzle at 100 feet per second, it can collapse part of a building atop a firefighter. Master streams, particularly improperly directed aerial straight streams, have collapsed brick chimneys, lifted roofs off wood buildings, and ex-

Figure 21.16. *Master streams deliver more than two tons of water at a velocity of 100 feet per second. If used incorrectly, master streams can kill or injure firefighters.*

ploded razor-sharp shingles and bricks from rooftops (Figure 21.17). As a general rule, interior firefighting should not be carried out in areas where powerful master streams are directed.

A mushrooming effect describes the horizontal flow at ceiling level and the subsequent banking down to floor level of smoke and heat generated by a fire in a confined space. The rapid "mushrooming" of smoke and heat traps and disorients firefighters during search and rescue operations (Figure 21.18). Mushrooming occurs more rapidly in small rooms. Venting roof skylights, stairways, and windows can delay or eliminate the mushrooming of smoke and heat in confined spaces during a fire.

Number 4 printed on an "identification of hazardous material diamond" (NFPA 704) can kill a firefighter if he does not know what it means. The number 4, printed in any one of the spaces of a hazardous

Figure 21.17. Master streams can collapse brick chimneys and sweep firefighters off rooftops.

Figure 21.18. Mushrooming of smoke occurs at top floors. Banked-down smoke can trap firefighters on the top floor.

material diamond, health hazard, flammability hazard, explosive hazard, or special information space, warns us that the hazard in the room or container is too dangerous to approach (Figure 21.19). Firefighters should immediately withdraw from the area and obtain expert advice about the hazard. There should be no attempt at firefighting.

Overhauling, the firefighting operation undertaken after a fire is under control, is designed to prevent the rekindle of a fire after the department leaves the scene. Firefighters have been killed and injured during the overhauling stage of a fire because of collapsing buildings, falls into open shaftways in darkness, carbon monoxide build-up in

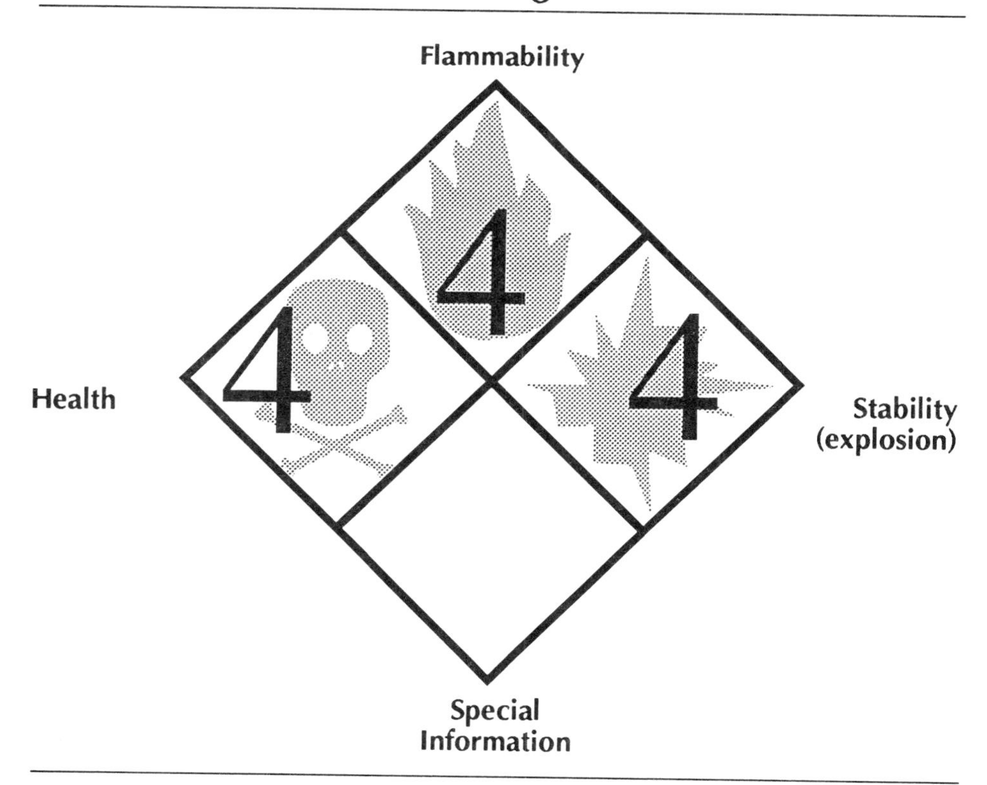

Figure 21.19. *The number 4, printed on an NFPA 704 diamond, "identification of hazardous material" placard, indicates a hazard too dangerous to approach with standard firefighting equipment.*

Figure 21.20. *Overhauling is an underestimated and dangerous phase of firefighting. From a health hazard point of view, overhauling is extremely threatening to firefighters.*

below-grade areas, electrocution, stress from the exertion of pulling down ceilings to examine the exposed spaces for hidden fire, and gas explosion. Overhauling is one of the most underestimated dangers of firefighters (Figure 21.20).

A parapet wall, the continuation of an exterior wall above the roof level, is the waist-high wall that encircles a roof. High decorative front parapet walls collapse suddenly during fires. There are three classifications of brick walls: free-standing walls, non-bearing walls, and bearing walls. A parapet wall is a free-standing wall. The stability of different types of wall during a fire is as follows (Figure 21.21):

Type of wall	*Stability during a fire*
Parapet wall	Least stable
Non-bearing wall	↕
Bearing wall	Most stable

Figure 21.21. *A parapet wall is a free-standing wall; it is less stable than a non-bearing wall or a bearing wall.*

A brick parapet wall extending over the large display windows of a one-story commercial building is supported by a steel I-beam. If the windows are vented during a fire, and flames flow out of the broken show windows, the heat can distort that steel I-beam. If the steel I-beam supporting the parapet wall distorts—sags, warps, or twists—it could cause the brick parapet wall above to collapse during the fire.

A peaked roof is a sloping roof supported at its highest point by a ridge rafter and at its lowest by a bearing wall. Examples of peaked roofs are gable roofs, mansard roofs, hip roofs, and gambrel roofs. A peaked roof is the most dangerous type of roof on which a firefighter can operate during a fire. In addition to the danger of collapse, a firefighter may lose his balance and fall off such a roof (Figure 21.22). The peaked roof is more dangerous than the flat roof because fire department ladders are required in order to gain access to the roof. There are no fixed exterior stairs leading to a peaked roof. Moreover, a peaked roof has a sloping surface, which makes walking and operat-

Figure 21.22. *A peaked roof is the most dangerous roof. A ladder is required to gain access to it, there is no enclosing parapet wall around it, and the surface is uneven and difficult for a firefighter to maintain balance when walking on it.*

ing on the roof more difficult. Peaked roofs do not possess parapet walls that could keep a firefighter from falling off. If a firefighter slips and loses his balance, he will roll off the roof; there is no parapet to stop his slide. Particular dangers of a peaked roof are that the rafters may collapse, the roof deck may collapse (even if the supporting rafters do not), and slate or tile shingles on a peaked roof may rain down on to firefighters operating at ground level.

Figure 21.23. Plastics burning generate large quantities of dense black smoke and have a rapid rate of flame spread.

Plastic is a material that contains one or more organic polymeric substances of large molecular weight. There are thousands of plastic products used as furnishings, fabrics, and building construction materials. Burning plastics are more dangerous to firefighters than burning wood, paper, or cloth. The flammability of the plastic product depends on its form; foam plastic products and foam fuzzy fabrics are the most flammable plastics. Burning foamed plastics create hotter fires. For example, a pound of polystyrene plastics can give off 18,000 British Thermal Units (BTUs) of heat, while wood or paper will only give off 7,000 to 8,000 BTUs of heat. Furthermore, the smoke given off by plastics is dense and black; the obscurity produced by plastic smoke at a fire is greater than that of smoke produced by wood or paper (Figure 21.23). Visibility in a smoke-filled room during a plastic fire can disappear suddenly and rapidly. In addition to the amount of heat emitted by burning plastic and the quick obscurity caused by the smoke of burning plastic, the rate of burning during a plastic fire is greater. The rapid rate at which plastic burns speeds up flashover inside a room. The environment in which a firefighter works—the burning room—has become more dangerous over the past 30 years, with the increased use of plastics as furniture, fabric, and building construction material.

The "point of no return" is the maximum distance that a fully equipped firefighter can crawl inside a superheated, smoke-filled room and still escape alive if a flashover occurs. The point of no return is five feet inside a doorway or window.

A reflash fire (also called a *flashback*) is the sudden ignition of flammable gases or smoke inside a smoldering, burned-out room where the initial fire has just been extinguished by a portable extinguisher or hose stream. After a fire has been "knocked down" and the hose stream shut off, there may still be sufficient heated gases and smoldering embers in a room to reflash suddenly if oxygen is admitted inside the area. Reflash fires often trap firefighters making a quick primary search after the fire has been extinguished. Such fires are often caused by burning foamed plastic mattresses and fires involving fuel oil burners in basements.

A residence building, specifically the one- and two-family type of house, is the occupancy where most fires occur and most firefighters are killed and injured. The three types of fires in which most firefighters are killed are:

Type of fire	*% of firefighters killed each year*
Residence building	28 of firefighter deaths
Wildfires	19 of firefighter deaths
Stores and offices	14 of firefighter deaths

Rollover is the sporadic ignition of combustible gases at ceiling level during the growth stage of a fire; it occurs before the room flashes over. *Rollover,* along with *high heat* and a *smoke level* banked down to half the height of a room in a residential building, are the warning signs of flashover (Figure 21.24). Firefighters should withdraw from a smoke-filled room when rollover starts to occur. Rollover will be

Figure 21.24. *Rollover is the sporadic ignition of combustible gases generated by fire. Rollover precedes flashover; it is a warning sign of flashover.*

visible near the ceiling level of a smoke-filled room or will be mixed with heat and smoke flowing out of the top portion of an open doorway or window.

Smoke consists of finely divided particles of soot and aerosols which accompany an uncontrolled fire. Smoke caused by incomplete combustion kills and injures firefighters (Figure 21.25) in the following ways: It causes reduced visibility, disorientation, entrapment, asphyxiation, explosions, and flashover. To reduce the dangers of smoke during a fire, ventilate the smoke-filled area in a coordinated, controlled manner.

Smoke explosions are caused by the random accumulation of combustible, smoke-filled atmospheres in confined spaces during a structural fire. Smoke explosions often occur in the main fire area during the growth stage of a fire and in its decay stage (Figure 21.26). When a smoke explosion occurs in the main fire area during the decay stage of a fire, it is sometimes called a "backdraft explosion." A smoke explosion can occur in a room adjoining a fire during the fire's fully developed stage. For example, a smoke explosion could occur in a smoke-filled room on either side or above the room that is actually

Figure 21.25. *Smoke kills most civilians and firefighters trapped in fire. Self-contained breathing apparatus must be worn by firefighters.*

burning; it happens when smoke seeping from the main fire into adjacent spaces creates a combustible atmosphere. When adjoining combustible rooms are opened up by searching firefighters, a smoke explosion occurs. The heat of the main fire area provides the ignition source; the fuel is the combustible smoke that has spread to the adjoining spaces and created a combustible atmosphere; the oxygen arrives with the initial entry of the firefighters searching for fire victims and fire spread.

Speed is the urgency with which firefighters are required to perform dangerous tasks in the vicinity of a life-threatening fire. Speed kills firefighters. Haste or the requirement to perform an act quickly

Figure 21.26. *Smoke explosions are caused by the random accumulation of combustible smoke and heat in confined spaces. Air is the trigger of a smoke explosion.*

before a spreading fire cuts off escape can cause a firefighter to make an error in judgment which results in his death or serious injury. Slow down! Pace yourself at a fire. Do not become caught up in the excitement of the fireground scene. Think of what you must accomplish at the fire, and do it. Don't let the fire dictate your actions. You should have had a pre-planned assignment before responding to the fire. Stick with the pre-plan and accomplish your assigned duty, even if others don't accomplish their assigned duties.

Stress is the physical and psychological exertion and pressure caused by the demands and dangers of firefighting (Figure 21.27). Stress from firefighting may cause cardiac arrest, stroke, or aneurysm (the rupture of the wall of an artery). Firefighters aged 46 to 51 are

Figure 21.27. Stress is caused by the physical and psychological exertions of firefighting. It is the leading cause of firefighters' deaths on the fireground.

those most often killed by the physical and psychological stress of firefighting.

The "tail board" (back step) or side step of a fire apparatus indicates the platform at the rear or side, respectively, of a fire truck. Firefighters riding on the tail board or side step of a responding fire apparatus are often killed during a collision, short stop, or sharp turn. One of the most dangerus actions a firefighter can take while responding to an alarm is to ride on the tail board or side step. Each year 25 percent of the annual death toll occurs to firefighters responding and returning to alarms; many of those fatalities involve riding or falling from the side or rear step of the apparatus.

A truss is a structure composed of wooden or steel members joined together in a group of triangles, fastened by metal bolts, sheet metal surface fasteners, or welds. Truss construction provides a dangerous roof or floor design when exposed by fire. The large surface-to-

Figure 21.28. Timber trusses and lightweight wood trusses collapsing during fires have killed 21 firefighters over the past three decades.

mass ratio of the many small interconnecting truss members make the structure vulnerable to early collapse. Over the past two decades, wood truss roofs have killed 21 firefighters by collapse (Figure 21.28). Truss roofs kill firefighters working below the truss, on top of the truss, and outside the truss-roofed building. When a timber truss roof collapses, it can cause the collapse of an outside bearing wall.

Tunnel vision is a psychological visual phenomenon that is experienced by firefighters during stressful firefighting situations. While focusing narrowly on a spectacular or dangerous event, the firefighter blocks out the nearby surrounding hazard or deadly peril. By concentrating on one point of the fire and not sizing up the entire fire area, a firefighter may block out an approaching danger. Tunnel vision and the hurried pace of firefighting cause accidents that could be avoided by a size-up of the entire fire and a slower pace.

An uncontrolled environment describes the dangerous, smoke-filled, collapse-prone, explosive atmosphere in which a firefighter works (Figure 21.29). Only combat soldiers and firefighters must perform this duty in such a dangerously uncontrolled environment. Even

Figure 21.29. *The uncontrolled environment in which firefighters work is the cause of many deaths and injuries.*

coal miners must, by law, be provided with lighting, fresh air, and structural supports in the mine before they go to work. Firefighters, when they crawl into a smoke-filled room, have no such safety guarantees. The poisonous smoke-filled room could explode, the building suddenly collapse, or the area burst into flame. Firefighters must bring their safety equipment with them: flashlights, protective breathing equipment, and powerful hose streams. Firefighters are the only workers in America who work in such a dangerous environment: the burning building.

Visibility reduction due to smoke and darkness at a fire is a major contributing cause of fireground death and injury (Figure 21.30). All

Figure 21.30. RIGHT: *Visibility reduction due to smoke and darkness is a major indirect contributing cause of firefighter death and injury.*

Figure 21.31. *The downwind or lee side of a fire is the most dangerous position on the fireground.*

firefighters should carry personal flashlights; all fire departments should use spotlights and floodlights to increase safety on the fireground by providing improved visibility at night. Firefighters should be trained to operate in areas of reduced visibility; training exercises simulating a smoke-filled room should be given to all firefighters. Mask facepieces with eye lenses blacked out or covered can give the firefighters some idea of how to operate in a smoke-filled room with its reduced visibility. Poor visibility at fires is caused by smoke, darkness, mask facepieces, tunnel vision, and the interior of a structure to which electricity has been cut off because of the fire.

Winds that suddenly changed direction or gusted have killed and injured firefighters. A sudden gust of wind can cause a wildfire to flare up and trap a firefighter who is operating in high brush. A sudden change of wind direction that blows into a flaming window can drive fire and heat into the path of advancing firefighters who are searching for victims or operating an attack hose line. High winds can cause a treetop, "crown" fire to spread over the heads of firefighters operating in woods. Firefighters should always attempt to take advantage of wind direction (Figure 21.31); the safest position for them is the upwind or windward side. If it becomes necessary to cut off a wind-driven fire, do so by attacking from its flanks (sides). Do not attempt to attack a windblown fire head on.

Lessons to Be Learned

There are no new lessons to be learned from a firefighter's death or injury. The cause of a tragedy is usually an old lesson we have not learned or have forgotten along the way.

APPENDIX

Is Your Fire Department Safe?

NFPA 1500
A STANDARD FIRE DEPARTMENT OCCUPATIONAL SAFETY AND HEALTH PROGRAM

NFPA 1500 is the first comprehensive written document created by a committee of fire service experts assembled by the National Fire Protection Association. This standard's purpose is to reduce the annual fire service death and injury toll. Each year 100 to 130 firefighters are killed and another 100,000 injured.

The following are ten important safety and health recommendations set forth by NFPA 1500, the standard for fire department occupational safety and health program. For the complete standard, write to:

National Fire Protection Association
Batterymarch Park
Quincy, MA 02269

1. A fire department should adopt this NFPA standard 1500 and provide its safety procedures to firefighters involved in fire suppression, search and rescue, and other related activities.

   Does your fire department have such a plan? Yes ______ No ______

2. A fire department should appoint a designated fire department safety officer.

   Does your fire department have such an officer? Yes _______ No _______

3. A fire department should have a standard operating procedure, in writing, for all fire and emergency scene operations.

   Does your fire department have such an S.O.P.? Yes _______ No _______

4. A fire department should have fire apparatus which provide seats, seat belts, or safety harnesses for every firefighter riding the vehicle.

   Does your fire department have such fire apparatus? Yes _______ No _______

5. A fire department shall provide each member with protective clothing and self-contained breathing apparatus.

   Does your fire department provide such fire equipment? Yes _______ No _______

6. A fire department should have an incident command system to provide safe command and control during fires and emergencies.

   Does your fire department have such a command system? Yes _______ No _______

7. A fire department shall require all firehouses to comply with local safety codes, building codes, and health codes.

   Does your fire department provide safe firehouses? Yes _______ No _______

8. A fire department shall have all firefighters who engage in fire and emergency operations re-examined by a physician each year.

   Does your fire department provide such medical re-examinations? Yes _______ No _______

9. A fire department shall provide a member assistance program that identifies and assists members with stress and personal problems.

Does your fire department provide such a program? Yes ______ No ______

10. A fire department shall have a fireground procedure to be followed for the safe exit from a dangerous area or building in the event of a sudden dangerous change in fire conditions.

Does your fire department have such a safe exit fireground procedure? Yes ______ No ______

Index